The Erotic Life of Manuscripts

The Erotic Life of Manuscripts

New Testament Textual Criticism and the Biological Sciences

YII-JAN LIN

OXFORD
UNIVERSITY PRESS

OXFORD
UNIVERSITY PRESS

Oxford University Press is a department of the University of Oxford.
It furthers the University's objective of excellence in research, scholarship,
and education by publishing worldwide. Oxford is a registered trade mark
of Oxford University Press in the UK and in certain other countries

Published in the United States of America by Oxford University Press
198 Madison Avenue, New York, NY 10016, United States of America

Library of Congress Cataloging-in-Publication Data
Lin, Yii-Jan, author.
The erotic life of manuscripts : New Testament textual criticism and the
biological sciences / Yii-Jan Lin.
p. cm.
Includes bibliographical references and index.
ISBN 978–0–19–027980–6 (hardback)
1. Bible. New Testament—Criticism, Textual. 2. Metaphor in the Bible.
3. Biology—Miscellanea. 4. Bible. New Testament—Manuscripts. I. Title.
BS2525.L56 2016
225.6—dc23
2015014756

3 5 7 9 8 6 4 2

Printed by Edwards Brothers, USA

Dedicated to my wai gong,
Mou-Chin Chen,
in his 100th year.

寧静致遠，順其自然

Contents

Acknowledgments

THIS BOOK BEGAN as a dissertation at Yale University, where I had the blessing (albeit at times a blessing in disguise) of having an adviser who detests being bored or subjected to arguments based on smoke and mirrors—and is forthright in saying so. I am grateful to Dale Martin for encouraging creativity and audacity in scholarship from the start of my time at Yale, for never being stingy with praise, and for his ability to cut to the heart of the matter with a few simple questions. I am indebted to his going to bat for me time and again, and I hope this work partially repays the faith he has shown in my scholarship.

I would also like to thank the reading committee of the dissertation, Harold Attridge and Adela Yarbro Collins, for their support and suggestions during writing, as well as Bart Ehrman and Eldon Epp for taking the time to consider an outsider novice's work on the field of their expertise. I am particularly grateful to Eldon Epp for reading drafts of several chapters ahead of time, offering advice, and generously sending offprints and suggesting further resources for research. Without Bart Ehrman's guidance and support, this work would still be mired in the process of finding a publisher. I would also like to thank David Parker for looking over chapter drafts and offering his encouragement from when I first presented the ideas of this project at the 2011 Society of Biblical Literature Annual Meeting in San Francisco.

A travel grant from the Yale Department of Religious Studies allowed me to travel to Münster and visit the Insitut für Neutestamentlische Textforschung. It was there that Gerd Mink and Klaus Wachtel graciously allowed me an interview and showed me the ins and outs of CBGM software. I thank them for permitting a transcription of that interview to be included in this book.

I was honored to receive generous funding from the Mrs. Giles Whiting Foundation for this project, and I am also grateful to the

Baden-Württemberg Exchange for making possible a year of study at the Ruprecht-Karls Universität in Heidelberg. The groundwork for much of this project was laid out during long solitary walks on the Heidelberg Philosophenweg. I must mention Bjørn Paarman, who formed part of my Heidelberg family (and continues as family from afar), and also Ted Lendon, whose interested skepticism helped shape the course of my thoughts.

I am grateful to David Horrell, Siam Bhayro, and the Exeter Centre for Biblical Studies for inviting me to share my ideas with them during my stay in Exeter. John Yieh, Diane Chen, and the entire Ethnic Chinese Biblical Colloquium also granted me the honor of presenting at our meeting in Baltimore in 2013, and I am especially grateful to my respondents, Wayne Kannaday and Tat-siong Benny Liew, for their marvelously helpful comments and critiques. Benny deserves special thanks as a behind-the-scenes mentor and advocate who has opened doors and given me opportunities from the beginning, when I first stumbled into an ECBC spring meeting in Chicago in 2006.

While a visiting instructor at the Pacific School of Religion, I was generously granted a light course load and no administrative duties so that I could finish this work. Since taking on a full course load and administrative tasks, I can still say that the faculty and staff at PSR are an inspiration and encouragement to my scholarship. I especially appreciate the faculty's suggestions and insights at the presentation of this project at our research seminar.

I must also thank Cynthia Read, my editor at Oxford University Press, for her enthusiasm and support from the very beginning of working together, and I am also grateful to Glenn Ramirez for guiding me through my first book publication.

I am humbled by the steadfast loyalty and intellectual depth of the friends who have buoyed me up in the process of writing. The friendships of my Yale cohort, Jeremy Bakker, Christina Harker, Dylan Burns, Yonatan Moss, Devin Singh, and Marko Geslani, have been invaluable. I am especially indebted to Jeremy, who has been my guardian angel from our first coffee in New Haven through the final throes of writing. The hospitality and genuine camaraderie of Jessica McCutcheon and Benjamin Purzycki, along with the very practical loan of books and a voice recorder, saw me through my last year of residence at Yale. And to Michael Thate I offer my best thanks for commenting on chapters, for exploring countless avenues of thought with me, and for his incorrigible friendship and enthusiasm.

Finally, this project would have been impossible without my family, immediate and extended, whose patience, encouragement, support, and endless help have bolstered me throughout this entire process, through many different trials. For my parents and my sister there really are no words of thanks that are adequate. I have been blessed with a son who has blithely cooperated with travel and changes, who does not deserve to be called "Grumpus" but fills my life with shrieks of laughter, mine and his. But for John Noël Dillon, philologist and historian, who fully deserves his nickname *Der Kleine Pauly*, this work would have been lesser in every way.

The Erotic Life of Manuscripts

Introduction

We have, on the one hand, imagination, synthetic thought, gift exchange, use value, and gift increase, all of which are linked by a common element of eros, or relationship, bonding, "shaping into one." And we have, on the other hand, analytic or dialectical thought, self-reflection, logic, market exchange, exchange value, and interest on loans, all of which share a touch of logos, of differentiating into parts.

LEWIS HYDE, *The Gift*

καὶ ὁ λόγος σὰρξ ἐγένετο καὶ ἐσκήνωσεν ἐν ἡμῖν

And the Word became flesh and dwelled among us.

JOHN 1:14

I BORROW THE title of this present work, *The Erotic Life of Manuscripts*, from writer and cultural critic Lewis Hyde and his book, *The Gift: Imagination and the Erotic Life of Property*. In that work, Hyde juxtaposes gift economics, which he associates with *eros*, and market economies, associated with *logos*. A market economy operates on calculation and logic, impersonal debt and payment; it is an environment in which goods are assigned a market value, and sold without requiring any personal relationship.[1] In a gift economy, property can be said to have an erotic life because when it

1 The epigraph comes from Lewis Hyde, *The Gift: Imagination and the Erotic Life of Property* (New York: Random House, 1979), 155. On the market economy, see ibid., 60–1: "I would begin the analysis by saying that a commodity has value and a gift does not. A gift has worth. I'm obviously using these terms in a particular sense. I mean 'worth' to refer to those things we prize and yet say 'you can't put a price on it.' We derive value, on the other hand, from the comparison of one thing with another. 'I cannot express the value of linen in terms of linen,' says Marx in the classic analysis of commodities which opens his *Capital*. . . . It is characteristic that commodities move between two independent spheres. We might best picture the difference between gifts and commodities in this regard by imagining two territories separated by a boundary. A gift, when it moves across the boundary, either stops being a gift or else abolishes the boundary. A commodity can cross the line

is given as a gift it can circulate, create bonds, and reproduce and multiply through the relationships and community it creates.

Hyde uses the ceremonial copper plaques that once circulated among the indigenous peoples of the Pacific Northwest to illustrate the circulation and increase of a gift as it is exchanged.[2] During a potlatch given by one tribe for another, the hosting tribe would present a copper as a gift to their guests, and the guests would, in turn, present blankets, food, fuel, and other things as part of the gift exchange. With each exchange of the copper between tribes, the return gift for the copper would increase, and the status and good relationship of the tribes depended on the circulation of coppers. A copper could be broken into pieces at a funerary potlatch for a chief, and the pieces would circulate having increased value, symbolic of the chief's spirit that continues and multiplies through the lives of the tribe members:

> In the case of the mortuary potlach, a material thing symbolizes a biological fact, the survival of the group despite the death of the individual. But it may be that the group would not survive as a group (and individual life would not survive, then, either) if these "biological" facts could not be expressed symbolically. We are social and spiritual beings; at some level biological, social, and spiritual life cannot be differentiated.[3]

In a way similar to the coppers of the tribal potlatch, the ancient manuscripts of the New Testament also operated in a gift economy, having erotic lives of their own. Up through the fourth and fifth centuries, the production, acquisition, and circulation of manuscripts operated mainly through social networks: a church leader might request a copy from another leader, or an author might forward copies of his own works to a circle of private friends for dissemination—or "publication."[4] There existed a

---

without any change in its nature; moreover, its exchange will often establish a boundary where none previously existed (as, for example, in the sale of a necessity to a friend). *Logos*-trade draws the boundary, *eros*-trade erases it."

2 Ibid., 28–36.

3 Ibid., 33–4, note. Hyde uses the alternative spelling, "potlach."

4 See Kim Haines-Eitzen, *Guardians of Letters: Literacy, Power, and the Transmitters of Early Christian Literature* (Oxford: Oxford University Press, 2000), 77–104 on the transmission and circulation of Christian texts in the second and third centuries; Harry Y. Gamble, *Books and Readers in the Early Church: A History of Early Christian Texts* (New Haven, CT: Yale University Press, 1995), 82–143 on the publication and circulation of Christian literature

book trade in the ancient world, as well as bookshops, and libraries, but the movement of books occurred mainly along the lines of personal relationships.[5] As William Harris states in *Ancient Literacy*: "the primary way of distributing books was not in any case by means of a trade of any kind but through gifts and loans among friends."[6]

Moreover, in a far more literal way than the coppers, the "gifts and loans" of manuscripts indeed grew and increased: one major reason for acquiring a text of the New Testament, other than for use—private or ecclesial—was to copy it. The exchange and circulation of manuscripts through friendships and communal relationships, ecclesial networks and scholarly requests for favors, meant the reproduction and multiplication of manuscripts. So, in a manner quite like the potlatch coppers, a manuscript is "a material thing that symbolizes a biological fact,"[7] that is, ancient manuscript reproduction and growth in numbers reflect the growth (through biological reproduction and spiritual conversion) of early Christian movements. In fact, these New Testament texts may be seen as reproductive in at least two ways: first, copies and copies of copies are produced physically throughout the ancient world; second, the texts themselves, their spiritual content and message, help create converts that grow movements, which in turn creates a greater demand for copies of the text. These phenomena of church life, faith, message, and textual transmission reinforce the idea that "at some level, biological, social, and spiritual life cannot be differentiated."[8]

---

through the fifth century; also Leighton D. Reynolds and Nigel G. Wilson, *Scribes and Scholars: A Guide to the Transmission of Greek and Latin Literature*, 3rd ed. (Oxford: Clarendon Press, 1991), 1–29 on publication during the Roman Republic and under the Early Empire, especially 24: "From the exchange of letters between Cicero and Atticus we can get a good idea of the casual and fluid nature of publication in the ancient world. There was no copyright or royalty (hence the importance of literary patronage) and private circulation could easily pass by degrees into full-scale publication; an author was able to incorporate changes into a text he had already published by asking his friends to alter their copies, but other copies [i.e., copies of those copies] would remain unaltered." The use of the masculine pronoun in the main text is intentional, reflecting probable, though unfortunate, historical realities.

5 Martial is an early example of a popular poet whom booksellers published for profit (though his works were also subject to circulation via various personal connections). See John Patrick Sullivan, *Martial: The Unexpected Classic* (Cambridge: Cambridge University Press, 1991), 15 with n. 32.

6 William H. Harris, *Ancient Literacy* (Cambridge, MA: Harvard University Press, 1989), 225. Quoted in Haines-Eitzen, *Guardians of Letters*, 82.

7 Hyde, *The Gift*, 33.

8 Ibid., 34.

Even after the fifth century, when centers for scribal work, or scriptoria, grew in number throughout Europe,[9] so that books could be copied in greater numbers and acquired through more impersonal, trade-driven networks, the personal, human element remained—and would remain until the invention of the printing press in the fifteenth century. Every manuscript before then had to pass through the hands of a scribe, its text had to be written in his script and subject to his skill, dedication, and physical endurance.[10] Of course, this meant that a manuscript was also subject to a scribe's deficits in any of these areas. If, for example, a scribe's attention wandered, or if he felt free to amend the text to fit his own theological viewpoints,[11] or even added his own illustrations or notes of complaint about the chilliness of the monastery,[12] then the manuscript would bear that mark of his personal touch.

In this way, within social networks and even the larger book trade that developed in the medieval period, the creation of manuscripts, their circulation, and eventual reproduction all involved a web of relationships

---

9 See Gamble, *Books and Readers*, 88 on "mass" production of manuscripts in scriptoria, achieved through large groups of scribes gathered to write by dictation: "Although the theory is attractive and commends itself to modern notions of efficiency, the evidence that such a scene actually took place before the early medieval period is poor."

10 On the physical challenges of scribal work, see Bruce M. Metzger and Bart D. Ehrman, *The Text of the New Testament: Its Transmission, Corruption, and Restoration*, 4th ed. (New York: Oxford University Press, 2005), 27–9.

11 See the classic work on intentional scribal changes, Bart D. Ehrman, *The Orthodox Corruption of Scripture: The Effect of Early Christological Controversies on the Text of the New Testament* (Oxford: Oxford University Press, 2003). Also, on a scribe's individual theological views affecting the text, see Haines-Eitzen, *Guardians of Letters*, 16: "That the scribes who copied early Christian texts were theologically invested in the texts they (re)produced, and worked from within private networks of friendships and acquaintances, illumines the intersections between texts and authority as well as literacy and power in early Christianity. . . . The work of early Christian scribes highlights the importance of texts in early Christianity and the ability to manipulate the very words of these texts when necessary; furthermore, such manipulations illustrate how scribes staged a certain power over the texts they copied." For an opposing view, see Gerd Mink, "Editing and Genealogical Studies: The New Testament," *Literary and Linguistic Computing* 15 (2000), 52: "Some essential assumptions are considered more probable than their contrary (without excluding that the contrary might happen at some points): (i) a scribe wants to copy a manuscript with fidelity; primarily the scribe does not want to create new readings; (ii) if the scribe introduces other readings, they come from another source (normally a manuscript); (iii) if the scribe uses more than one source, few rather than many sources will be used; and (iv) the source copies have closely related texts rather than less related ones." See also the discussion of CBGM in Chapter 4.

12 For an example of such informal marginalia, see Metzger and Ehrman, *Text of the New Testament*, 32.

among friends, authors, and patrons, between a scribe and his work, and between exemplars and their copies.[13] Textual characteristics left behind by one scribe in a manuscript that would serve as an exemplar for other copies would be passed on, marking generations of descendants, so to speak.

This metaphor of sexual reproduction, of "erotic" life, is exactly the one used by textual critics to describe the extant manuscripts and texts they study. In almost any handbook or monograph of textual criticism, manuscripts or texts are described as parents and offspring, comprising families, and mingling their textual makeup to create new mixtures and hybrids. Even the changes made by scribes as they create new copies can be cast in semi-erotic terms, fraught with danger: texts are "adulterated" and "corrupted," as the unfaithful scribe creates variant readings, perhaps yielding to the temptation to make scripture conform to his own theological views.[14]

This study began as an investigation of the use of these metaphors in textual criticism. Let me preface this discussion before I go any further by stating that I am not a textual critic, nor do I attempt to do the work of a textual critic in this book. My interest in the specific terminology of textual critics came from observing the arguments made by scholars on either side of what might be called "The Great Divide" in biblical studies. On the one side stand scholars of literary criticism, particularly poststructuralist and postcolonial criticism and minoritized hermeneutics. On the other stand scholars whose methodologies are squarely historical-critical, including those involved in the more "technical" aspects of the field, such as archeology, papyrology, and textual criticism or, as it is also called, "lower criticism."

It seems that literary theorists and critics are accusing historical-critical scholars of being only positivistic and empiricist, mired in traditional, Western ways of thinking. Historical-critical scholars, on the other

13 Some of this might still be said of the authoring and publication of books today, although the intermediary element of computers and computer editing, along with a much more massive scale of distribution and marketing, distances author, editor, and reader from one another to a great degree.

14 See, for example, Gamble, *Books and Readers*, 123: "Complaints about the *adulteration* of texts are fairly frequent in Christian literature. . . . In some respects [Christian texts] were more *vulnerable* than ordinary texts . . . editors and copyists must often have revised texts in accordance with their own perceptions. This *temptation* was stronger in connection with religious or philosophical texts than others simply because more was at stake." Emphasis added. In Gamble's description, the greater the sanctity of the text, the greater the vulnerability that text has to being adulterated.

hand, accuse the other side of following passing trends while ignoring historical or "diachronic" analyses of the text. A good example of the latter critique is provided in John Ashton's chapter on "narrative criticism" in his book *Studying John: Approaches to the Fourth Gospel.* He begins the chapter provocatively enough:

> the historical critical method . . . is still alive and running strongly. In the last decade or so, however, it has had to compete with a new and very different kind of approach . . . I shall use the label "narrative criticism": it is distinguished from earlier fashions by a consistent vision of the Gospels as, above all, *stories.* I shall argue that narrative criticism is more of a fad than a fashion and that since it misconceives the true nature of the Gospels the results it yields are trifling, if not altogether illusory.[15]

There are a few significant points to note in this critique. First of all, the historical-critical method and narrative criticism are understood to be in *competition* with each other, with historical criticism being very much in a defensive position. Second, Ashton assumes that there exists one "true nature of the Gospels." Finally, narrative criticism and its ilk are designated as *fads* rather than true, scholarly methods.

A representative from the other side of the great divide is Fernando Segovia, whom Ashton happens to critique in the chapter just quoted. Segovia, in *Decolonizing Biblical Scholarship,* describes historical criticism as having "little conception of the text as a literary, strategic, and ideological whole," being built on a "strong positivistic foundation" and regarding the text as "univocal and objective."[16] Although Segovia writes in a far more diplomatic fashion than Ashton (for example, he states that historical criticism is a perfectly viable method capable of interaction with literary and cultural criticism), Segovia implies, by the use of the term *decolonizing,* that something more insidious is at work in traditional methods. This especially seems the case when Segovia proposes a cultural analysis that takes into account "the shadow of the empire in the production of modern readings of the ancient texts."[17]

---

15 John Ashton, *Studying John: Approaches to the Fourth Gospel* (Oxford: Oxford University Press, 1994), 141. Emphasis original.

16 Fernando Segovia, *Decolonizing Biblical Scholarship: A View from the Margins* (London: Orbis, 2000), 13.

17 Ibid., 126.

It seems that there is no common ground between these two approaches to biblical studies. However, as I surveyed the methods of "diachronic" studies of the gospels, especially textual criticism, a language used by both these methods and those of literary theorists began to emerge: the language of family ties, sexual reproduction, genealogy, in short, an "erotic" language of relationships pure and impure, productive, dangerous, and ambivalent. This language in textual criticism is further complicated by *racial* terms, to speak of difference, mixture, and corruption. Here are just a few examples from some very important works of the field:

> texts are not independent: they are all fragments of *a genealogical tree* of transmission.[18]
>
> . . . once manuscript categories have been established on the basis of textual *consanguinity*, they can be evaluated qua groups.[19]
>
> The question therefore is whether we have a unified tradition or whether the tradition is *hybrid*.[20]
>
> Only variations . . . could be used when establishing the *physiognomy* of the other text-types.[21]
>
> . . . it is possible that the lost papyri outside of Egypt would display a drastically different textual *complexion*.[22]

The italicized terms above, often used in textual criticism (not to mention "original," "ancestral," and "corruption," among others), are all involved in heated discussions in race and ethnic theory, postcolonial theory and ethnic minority hermeneutics. How, then, do they function in textual criticism? Where did this discourse come from and what are the limits of its utility?

---

18 Fenton John Anthony Hort, "Introduction," in *The New Testament in the Original Greek*, ed. Brooke Foss Westcott and Fenton John Anthony Hort (Cambridge: Macmillan, 1882), 39–40. Emphasis added.

19 Metzger and Ehrman, *Text of the New Testament*, 232. Emphasis added.

20 David C. Parker, *An Introduction to the New Testament Manuscripts and Their Texts* (Cambridge: Cambridge University Press, 2008), 280. Emphasis added.

21 Bart D. Ehrman, "Methodological Developments in the Analysis and Classification of New Testament Documentary Evidence," in *Studies in the Textual Criticism of the New Testament*, New Testament Tools and Studies 33 (Leiden: Brill, 2006), 15. Emphasis added.

22 James R. Royce, *Scribal Habits in Early Greek New Testament Papyri*, New Testament Tools, Studies and Documents 36 (Leiden: Brill, 2008), 16. Emphasis added.

These are the questions that began this project, and it pursues two main goals. The first is methodological: to see if cultural studies could operate within biblical studies in a way that does not "merely" read the Gospels (or Paul or other texts) as *"stories"*—as Ashton accuses it of doing—and then retell them. Instead, I aim to analyze and deconstruct a technical aspect of the field, textual criticism, a subfield with which cultural studies usually has nothing to do. The second goal is to explore the biological metaphor and its surrounding discourse used by textual critics to see how it has shaped the field, its goals and assumptions, and illuminate the cultural contexts, intellectual milieu, and set of preoccupations and anxieties from which it arose.

The structure of this book is therefore chronological, following the developments of New Testament textual criticism as it borrowed from the discoveries and methods of the natural sciences, beginning in the eighteenth century and continuing to the present. Heretofore, numerous histories of the methods of New Testament textual criticism have been written, some beginning the narrative as far back as Origen and Jerome. But these have largely been emic studies that take for granted the use of taxonomy and genealogies in a search for origins and authenticity. Even more recent discussions, which question the stability of the "original text" and its discovery as *the* goal of textual criticism, do not consider how terminology, structures, and metaphors have brought the field to its present state. This project undertakes a Foucauldian analysis of the history of textual criticism and its discursive formation, focusing on the shared concerns of New Testament textual criticism and the biological sciences with classification, origins, contamination, genealogy, and evolution.

The chapters are divided into two larger sections. The first section, comprising Chapters 1 and 2, is titled "Collection and Theorization," which covers the eighteenth and nineteenth centuries, a period of New Testament textual criticism full of the discovery, collection, and organization of ancient manuscripts. It was also a time in which modern textual critics formed the foundations of philological methodology: the principles of evaluating texts, establishing the relationships between readings, and the canons of internal evidence. The second section, "Historicization and Innovation," covers the twentieth century to the present, the period of history discussed in Chapters 3 and 4. This section focuses on the interpretation of the transmission of the New Testament as a historical *narrative*, rather than a stark genealogy of corruption, and the use of new statistical methods and computer software to map comprehensive stemmata of New Testament texts.

The first chapter, "Bengel and the Classification and Racialization of Texts," begins with the start of modern textual criticism, roughly coinciding with the work of Johann Albrecht Bengel (1687–1752). Bengel's greatest innovation and contribution to textual criticism was his division of extant manuscripts into "families, tribes, nations," so that a philologist need not deal with every individual manuscript but could evaluate groups of texts classified together because of their similar readings. The two "nations" into which Bengel categorized manuscripts were the Asiatic and the African, and although these names were given primarily to designate the geographical provenance of manuscripts, Bengel also described texts and readings in specifically racial terms. This impulse to classify and racialize ran parallel to developments in the natural sciences. Carl Linnaeus (1707–1778) developed a systematic and comprehensive taxonomy of all biological life that remains in use today. Linnaeus divided the human species into four races, differentiated by geography and skin color, and other Enlightenment thinkers and scientists, such as Kant and Buffon, would categorize and recategorize human races. Textual critics of the eighteenth and nineteenth centuries likewise classified and reclassified text families, describing these families or nations in distinctively racial terms.

With a system of classification in place, the differences between categories and their relationships with one another became the next preoccupation of both textual critics and natural scientists. Chapter 2, "Lachmann and the Genealogy and Corruption of Texts," discusses how the classification of texts and, consequently, their hierarchization as reliable or unreliable witnesses, led to a genealogical understanding of textual transmission. Karl Lachmann (1793–1851) effectively introduced the genealogical method to New Testament textual criticism by insisting that scholars adjudicate among text families by taking into account the history of textual transmission, imagined as a genealogy of texts, with an ancestral archetype at the top of the family tree. This mirrored the hierarchization of the human races by natural scientists, with the "Caucasian" or "White" race as the progenitor of all other races, and from which issued a genealogy of racial degeneration and corruption.

Lachmann's edition of the Greek New Testament was the first to depart completely from the *Textus Receptus* (originally Erasmus's edition from the sixteenth century), and both this freedom from reverting to the *Textus Receptus* as the very Word of God and the genealogical method created a new ultimate goal for New Testament textual criticism: the "original text," the pure, unsullied ancestor. The variant readings descended from the original were, like lesser races descended from the original Caucasian race, degenerate and corrupt.

The concern over degeneration from an archetypal ancestor began to change in scientific circles toward the end of the nineteenth century as Darwin's theories of evolution and natural selection, presented in the publication of *The Origin of Species* in 1859, came more and more to be accepted. This seismic shift in understanding of the history of life on Earth eroded the concept of static, essentialist taxonomic categories and reconceptualized species as variable entities constantly in flux as they adapt to new environments. Chapter 3, "Darwin, Streeter, and Narrative Textual Criticism," traces the gradual shift in focus in New Testament textual criticism from classification and genealogy toward a Darwinian understanding of texts as constantly evolving and adapting.

This shift began primarily with B. H. Streeter (1874–1937) and his theory of local texts in *The Four Gospels: A Study of Origins*, which hypothesized different narratives for the development of texts in their locales over time. Much as a species introduced to a new environment adapts to a different climate, competition for resources, and predators, so a local text, in Streeter's theory, took on its own characteristic features as it was copied and corrected in its local context to reflect the resources and aesthetic or theological bent of its community. This new understanding of variants as part of the evolution of the New Testament text and not merely as corruptions was taken up by narrative textual critics in the last quarter of the twentieth century, and the goals of the field came to include not only searching for an "original text" but also examining the historical, sociological, and theological contexts of variants in the history of textual traditions.

Narrative textual criticism of the New Testament is part of a broader movement in textual criticism in general, sometimes called the New Philology. Its counterpart—and counterpoint—may be termed the New Phylogeny, "phylogeny" being the branch of science that maps the evolutionary relationships between organisms through time. The New Phylogeny maintains the traditional goal of textual criticism, namely the determination of the ancestral text, but uses the methodologies of phylogenetics and genomics, to arrive at it. Chapter 4, "Philology and Phylogeny," surveys the developments and methodologies of evolutionary biology and textual critics' explicit use of them, specifically in The Canterbury Tales Project, which is dedicated to mapping the evolutionary and genetic relationships of all of Chaucer's extant manuscripts.

The method in New Testament textual criticism most comparable to the New Phylogeny is the Coherence-Based Genealogical Method

(CBGM), used at the Institute for New Testament Textual Research in the production of the *Editio Critica Maior* and the Nestle-Aland *Novum Testamentum Graece*. The CBGM (and its software) allows textual critics to compare texts with one another at *every* point of variation in order to determine their relatedness. Because of this comprehensive approach, the older classification of text families is no longer relevant, just as the Linnaean taxonomy is disregarded in phylogeny. In stark contrast to narrative textual criticism, and similar to phylogenetics, the CBGM does not attempt to sketch any history, but focuses only on the mapping of textual relatedness.

Included as an appendix is the transcription of an interview I conducted with Academic Director Gerd Mink and Klaus Wachtel, coeditor of the *Editio Critica Maior*, at the Institute for New Testament Textual Research in Münster. Our discussion of the CBGM covers its overall methodological approach as well as Mink and Wachtel's positions on textual meaning, scribal practice, and narrative textual criticism.

It is clear that since the beginning of modern textual criticism, New Testament textual critics have used the discourse and methodologies of the biological sciences. It is not my goal to discuss here why they did, or whether they did so intentionally. Instead, to borrow from Foucault's terminology and concepts in *The Order of Things*, I take it for granted that textual critics generally operate—as all scholars do—within the *episteme* of their time, within and upon a certain space and topography of knowledge.[23] The three different *epistemes* outlined by Foucault, of the Renaissance, the Classical period, and the modern period, can be sensed impressionistically in the stages of New Testament textual criticism I discuss. For example, Foucault demonstrates the essential difference between the Renaissance and Classical periods thus:

> To the Renaissance, the strangeness of animals was a spectacle: it was featured in fairs, in tournaments, in fictitious or real combats, in reconstitutions of legends in which the bestiary displayed its ageless fables. The natural history room and the garden, as

23 For a brief definition and explanation of *episteme*, see Michel Foucault, *The Order of Things: An Archaeology of the Human Sciences* (New York: Vintage Books, 1973), xxi–xxii: ". . . on what basis knowledge and theory became possible; within what space of order knowledge was constituted; on the basis of what historical a priori, and in what element of what positivity, ideas could appear, sciences be established, experience be reflected in philosophies, rationalities be formed, only, perhaps to dissolve and vanish soon afterwards."

> created in the Classical period, replace the circular procession of the "show" with the arrangement of things in a "table." What came surreptitiously into being between the age of the theatre and that of the catalogue was not the desire for knowledge, but a new way of connecting things both to the eye and to discourse.[24]

This difference is almost perfectly reflected in the description of various readings by the Humanist Joseph Scaliger as "monstrous progeny" in contrast to Bengel's proposal to sketch them out in "genealogical tables." For the Humanist, the identification of a text was inseparable from its hermeneutic, whereas for the Classical scholar, a text was to be identified solely with its essence apart from interpretation—*sine interpretatione,* as Lachmann would propose. Bengel's classification and genealogical tables belong to the taxonomic impulses of his time. So, also, the shift from a classificatory understanding of readings to an evolutionary one follows Foucault's hypothesis that the Classical animal, a categorizable "living being," changed dramatically to an "organism," and, what is more, an organism understood historically rather than as statically occupying a square in a taxonomic grid.

Nevertheless, my task here is not to reconstruct the history of New Testament textual criticism in Foucauldian terms, and parts of the history I cover simply do not conform to the epistemological stages proposed by Foucault. The New Phylogeny and CBGM, for example, certainly reflect the loss of subjectivity that Foucault described as characteristic of the modern period, a period in which "man" finds himself not as a separate and removed observer and organizer of knowledge but as part of the tables and grids he created, in fact, always already knowing and being known only through discourse and not independent of it. In the same way, the New Phylogeny and CBGM operate so that the subjectivity of the textual critic is, to a great extent, dissolved in the encoding and analysis of the computer mind. The comparison of DNA to texts, to language, demonstrates the disappearance of the human subject into the discourse of genomics, textual and biological, where an organism's past and future are already encoded and bound.

Yet the practitioners of CBGM also input their text-critical decisions as data, and these are based on the traditional canons of textual criticism,

---

24 Ibid., 131.

centuries if not millennia old. The Linnaean/Bengelian taxonomy of text families and races is rendered obsolete by the method, but a firm belief in language as transparently representational is implicit in the treatment of texts as "carriers of meaning," which can be disfigured and also repaired. Thus, hermeneutic and fable (for example, the traditional precept *lectio difficilior potior*), analogy, representation, and the dissolution of subjectivity are all present in the construction and use of CBGM today.

A totalizing epistemological history is therefore irrelevant at points, and instead I trace the use of a discourse by textual critics, sometimes explicit and highly intentional, in order to deconstruct the assumptions and methods of the field. It is not my aim, however, to invalidate any part of New Testament textual criticism or any of its methods. Deconstruction, in this project, is rather the probing of unquestioned metaphors and structures of knowledge. I take the definition of "deconstruction" from Gayatri Chakravorty Spivak:

> Deconstruction does not say there is no subject, there is no truth, there is no history. It simply questions the privileging of identity so that someone is believed to have the truth. It is not the exposure of error. It is constantly and persistently looking into how truths are produced.[25]

Therefore, at no point would I suggest, for example, that the search for the oldest attainable text of the New Testament is an invalid enterprise. Instead, the query posed here is if attaining that text is important to scholars beyond simply answering a historical question ("what is the earliest attainable text?"), why? If, for example, textual critics propose that the oldest text, or the chimerical autograph, is equal to the most authoritative text, why do they suppose that is the case? Why privilege origin rather than development? Observing this privileging, as David Parker so eloquently puts it, "one is struck by the fact that a belief in single authoritative texts accords to the Spirit a large role in the formation of Scripture, and almost none at all in the growth of the tradition."[26]

---

25 Gayatri Chakravorty Spivak, *The Spivak Reader: Selected Works of Gayatri Chakravorty Spivak*, ed. Donna Landry and Gerald MacLean (New York: Routledge, 1996), 27.

26 David C. Parker, *The Living Text of the Gospels* (Cambridge: Cambridge University Press, 1997), 211.

It is my contention that the post-Lachmannian search for the "original text" is intimately tied to the biological and specifically racial metaphors used by textual critics, following the methods of scientists that first created "races" by grouping humans according to representative features, organizing these groups hierarchically through an imagined genealogy of races, and defining difference from a mythical, pure ancestor as degeneration and corruption. In the same way, textual critics came to understand variant readings and "hybrid" texts—texts copied from more than one exemplar—as corruption and contamination. Only with the disruption brought by Darwin and evolutionary theory did the field begin to change and focus on the narrative of textual development.

Of course, the entire world of biblical studies was changing rapidly through the twentieth century, and it can easily also be argued that the advent and increasing growth of narrative textual criticism was due to a general movement toward cultural studies and literary theory and away from historical-critical methods of interpretation such as form criticism. Narrative textual critics have themselves indicated that postmodernity indeed informs their focus on variant readings and a history of textual diversity. So Parker describes narrative textual criticism as "an approach [that is] not untouched by postmodernism,"[27] and Eldon Epp states:

> It is clear to me that we . . . have moved beyond the "certainties" of historicism and even beyond the comforts of modernism to a recognition that the notion of "multiple truths" is more than a novel fad or a passing fancy. . . . I have come to recognize that multiple interpretations by an array of fellow human beings across the globe are, in numerous instances, each defensible and credible when seen within their varying intellectual and socio-cultural frameworks. Acknowledged or not, this is the environment in which all of us now work, and in some way we all must come to terms with this new world.[28]

A "coming to terms with this new world" has not been limited to New Testament textual criticism but has been reflected in the work of philologists

---

27 David C. Parker, review of *The Orthodox Corruption of Scripture: The Effect of Early Christological Controversies on the Text of the New Testament* by Bart D. Ehrman, *Journal of Theological Studies* 45 (1994): 704.

28 Eldon J. Epp, "Introduction: A Half-Century Adventure with New Testament Textual Criticism," in *Perspectives on New Testament Textual Criticism: Collected Essays, 1962–2004* (Atlanta: Society of Biblical Literature, 2005), xxxv.

of other literature as well. In describing the challenges of producing an edition of the medieval romance *Parzival*, Michael Stolz presents the New Philology as a driving force in reconsidering the issues of an authoritative text, since "Its tendency is to renounce the hierarchy of individual manuscript sources in the interest of the fundamentally variable, unstable status of medieval manuscript culture."[29] Jerome McGann, scholar of nineteenth- and twentieth-century literature, crystallizes the major tenet of the New Philology in this way:

> The textual condition's only immutable law is the law of change. It is a law, however, like all laws, that operates within certain limits. Every text enters the world under determinate sociohistorical conditions, and while these conditions may and should be variously defined and imagined, they establish the horizon within which the life histories of different texts can play themselves out. The law of change declares that these histories will exhibit a ceaseless process of textual development and mutation—a process which can only be arrested if all the textual transformations of a particular work fall into nonexistence. To study texts and textualities, then, we have to study these complex (and open-ended) histories of textual change and variance.[30]

Here again, however, the discourse of science creeps in: even given a postmodern, poststructuralist understanding of the instability of language, the New Philology follows the law of change, like the law of gravity or the laws of thermodynamics. The law of change is a law "like all laws," and therefore subject to the same logic as, say, Newtonian physics. Furthermore, under this rubric, this law of mutability, texts exhibit essentially the same reactions and adaptations as the biological organism in Darwinian evolution. There is "textual development and mutation," and these depend on a text's "determinate conditions." These continual transformations cease only when "all the textual transformations of a particular work fall into nonexistence," or, put differently, all variations of a textual species become extinct.

---

29 Michael Stolz, "New Philology and New Phylogeny: Aspects of a Critical Electronic Edition of Wolfram's *Parzival*," *Literary and Linguistic Computing* 18 (2003): 140–1.

30 Jerome McGann, *The Textual Condition* (Princeton: Princeton University Press, 1991), 9. See also id., *A Critique of Modern Textual Criticism* (Charlottesville, VA: University Press of Virginia, 1983).

The language of textual criticism, and thus its epistemological foundation, is inextricably intertwined with scientific, biological discourse, even as textual critics turn their attention toward the instability of texts and away from the purported stability of observation, *data*, and the evaluation of readings *sine interpretatione*. On the other hand, defenders of the traditional goals of textual criticism—to produce an authoritative, stable text as close to the "original text" or "author's intention" as possible—use scientific discourse, and textual criticism's long relationship with scientific methods, to justify the continuation of the field's older raison d'être.[31] For example, Siegfried Wenzel, scholar of Middle English, describes the main task of textual criticism in these terms:

> To these [interpretive] and a host of other contemporary critical questions and approaches to literature, philology continues to serve as a handmaiden, furnishing the material basis on which they must stand. Handmaidens are proverbially humble and modest. . . . At the same time, scholarship is not an absolute monarchy but a republic, in which the handmaiden, while doing her job of preparing the necessities of life—intelligible texts and tools for their understanding—will also remain constantly watchful and critical of the nobility. To order the disciplines devoted to the understanding of literary texts hierarchically, in the shape of a pyramid with paleography at the base and semiotics at the apex, is tempting but dangerous, because such a model allows the semiotician as well as the literary critic in the middle ranges to remain above and aloof from the concerns of philology. Not just an ancillary discipline, philology is an attitude of respect for the *datum*, for the facts of the texts and its contexts, which should be cultivated at all levels of our enterprise to understand and appraise.
>
> Philology thus holds not only a material value, in that it provides the raw materials for understanding, but equally a disciplinary one, by continuously demanding that the intellectual systems built by interpreters or theoreticians be tested against and anchored in the realities of the subject matter.[32]

---

31 But certainly not its oldest raison d'être. Establishing author's intention and the original text are goals of modern textual criticism, not ancient.

32 Siegfried Wenzel, "Reflections on (New) Philology," *Speculum* 65 (1990): 17–8.

Wenzel attempts to characterize philology as both the sine qua non of all literary disciplines and as operating wholly apart from interpretive methods, providing only "the raw materials" for literary critics and semioticians. The argument rests on a false modesty, for clearly the humble handmaiden is the only one who deals with "the realities of the subject matter," and, in her "respect for the *datum*," is able to approach what is unprocessed, objective, static, "raw," and observable, against which all "intellectual systems" that are "built" must be tested. The "nobility" cannot stand aloof from the handmaiden, but she operates on a scientific plane apart from interpretive enterprises.

In an even starker manner, New Testament textual critic and editor of the *Editio Critica Maior* Barbara Aland states:

> Is not textual criticism a historical-critical science by definition, which determines what is, and which precisely may not introduce subjective understanding, much less arbitrariness, in an improper manner into its decision-making process? The so-called "internal criteria" of textual criticism may be indispensible, but how often are decisions based on the internal criteria reversible, precisely because they are subjective! Therefore, at the Institute for New Testament Textual Research, we strive to emphasize the external criteria, i.e. to give an objective and rational basis for an appreciation of the value of the witnesses in themselves and in the overall view of all witnesses.[33]

Aland defines the field as a science, and she assumes that science does not deal with the subjective. The "internal criteria" mentioned refers to the analysis of readings to determine which came prior to the others based on considerations of what is omitted, added, interpolated, harmonized, and so on, in the text itself. These Aland relegates to the realm of interpretation,

33 Barbara Aland, "Welche Rolle spielen Textkritik und Textgeschichte für Verständnis des Neuen Testaments? Frühe Leserperspektiven," *New Testament Studies* 52 (2006), 304: "Ist nicht Textkritik eine historischkritische Wissenschaft per definitionem, die feststellt, was ist, und die gerade nicht in ungebührlichem Maß das subjektive Verstehen oder gar Willkür bei ihrer Entscheidungsfindung einbringen darf? Zwar ist auf das sog. innere Kriterium der Textkritik keineswegs zu verzichten, aber wie oft sind Entscheidungen nach dem inneren Kriterium umkehrbar, eben weil subjektiv. Wir bemühen uns daher im Institut für neutestamentliche Textforschung das äußere Kriterium zu stärken, d.h. die Einsicht in den Wert der Zeugen für sich und im Konspekt aller Zeugen, objektiv und rational zu begründen." My translation.

the subjective, and therefore understands them as unstable and reversible. "External criteria," for example, the statistical quantification of identical readings between texts or the dating of manuscripts, are, in contrast, "objective and rational." These demonstrate, for Aland, that textual criticism is strictly a *science*, and therefore not subject to the arbitrary or reversals.

As a science, textual criticism "determines what is" (*festellt was ist*). A stronger pronouncement of the stability, objective simplicity, and importance of the field could hardly be made. But the history of textual criticism and the biological sciences provided here demonstrates that such stability and objectivity is a myth. The shifts from taxonomic essentialism to evolutionary nominalism, from degenerative genealogies to an ever expanding phylogenetic tree of life, show the mutability of science and scientific discourse. These shifts are not only the result of new discoveries but also of new methods and perspectives that reflect both the disruption of an *episteme* and new socio-political concerns and anxieties. In the same way, textual criticism has changed not only due to the discovery of new manuscripts but also of new methods and structures of knowledge corresponding to the exigencies of the scientific discourse it adopts.

Again, this is not to say that, therefore, evolutionary narrative criticism or the phylogenetically tinged CBGM are somehow unsound because they did not appear sui generis on the philologist's horizon. This also does not invalidate the groupings and descriptions of text families even if they use the white supremacist language of race theory from the eighteenth and nineteenth centuries. I have no interest in a politically correct overhaul of the terminology of the field. What interests me is the protean figure of the biologized, living text of the New Testament and how it has both expanded and limited the imagination of textual critics.

This philological incarnate word, its family and race, its sexual liaisons, its adaptations, and its genetic code, is the main character of the following chapters.

PART I

Collection and Theorization

I

Bengel and the Classification and Racialization of Texts

If there are no cleavages in nature, it is evident that our classifications are not hers. Those which we form are purely nominal, and we should regard them as means relative to our needs and to the limitations of our knowledge. Intelligences higher than ours perhaps recognize between two individuals which we place in the same species more varieties than we discover between two individuals of widely separate genera. Thus these intelligences see in the scale of our world as many steps as there are individuals.

CHARLES BONNET, *The Contemplation of Nature*

WHAT BONNET SAYS, in the epigraph above, regarding nature (he was a pure nominalist in regards to the category of "species") should also be said regarding history and kept in mind especially when writing a history of ideas.[1] Any starting point is more or less arbitrary, and the story of textual critics and natural scientists begins here, necessarily, *in medias res*. By the time Linnaeus published the first edition of the *Systema Naturae* in 1735,

1 The epigraph is a translation from Arthur O. Lovejoy, *The Great Chain of Being: A Study of the History of an Idea* (Cambridge, MA: Harvard University Press, 1933), 231, based on the second edition of Bonnet's *Contemplation* of 1769: Charles Bonnet, *Contemplation de la Nature, Nouvelle édition, corrigée et considerablement augmentée* (Hamburg: Virchaux, 1782), book 1, chapter 11, p. 53: "Mais, si rien ne tranche dans la Nature, il est évident que nos distributions ne sont pas les siennes. Celles que nous formons sont purement nominales, et nous ne devons les regarder que comme des moyens relatifs à nos besoins et aux bornes de nos connoissances. Des Intelligences qui nous sont supérieures, écouvrent peut-être entre deux Individus que nous rangeons dans la même espece, plus de variétés que nous n'en découvrons entre deux Individus de genres éloignés. Ainsi ces Intelligences voient dans l'Echelle de notre Monde autant d'Echelons qu'il y a d'Individus." On the history of the concept of species, see John S. Wilkins, *Species: A History of the Idea* (Berkeley: University of California Press, 2009), 10.

the discussion of a classification of the natural world was already more than two thousand years old. Of course, Plato and Aristotle (who effectively began the conversation for the Western world) were concerned with the larger project of classifying all possible units of existence, but this included the biological world, and their philosophical enterprise continued to exert its influence on the natural sciences until the nineteenth century.

Plato's and Aristotle's divisions and categorizations would later be incorporated into Porphyry's comb (or tree), which classified individuals through a series of dichotomies (e.g., corporeal vs. incorporeal, animate vs. inanimate), and Neoplatonist understandings of species and genus. A more encyclopedic approach to varieties in nature is found in Pliny the Elder's *Historia naturalis* (which is itself heavily influenced by Theophrastus and Dioscorides)[2] in its accounts of plants and animals, minerals, and heavenly bodies. The herbals and bestiaries of the late Middle Ages, which listed plants and animals, respectively, for medicinal or moralizing purposes, show Pliny's influence. The philosophical debate regarding universals and individuals (whether universals exist outside of the mind in addition to individuals) raised by Porphyry was continued by Abelard and William of Ockham.[3] At the dawn of the Enlightenment, Locke asserts his own nominalist position by stating

> [*Genera, species,* and their essences] appear, upon a more wary survey, to be nothing else but an artifice of the understanding, for the easier signifying such collections of ideas as it should often have occasion to communicate by one general term; under which divers particulars, as far forth as they agreed to that abstract idea, might be comprehended. And if the doubtful signification of the word *species* may make it sound harsh to some, that I say the species of mixed modes are "made by understanding"; yet, I think, it can by nobody be denied that it is the mind makes those abstract complex ideas to which specific names are given. And if it be true,

2 Erik Nordenskiöld, *The History of Biology: A Survey*, trans. L. Eyre (London: Kegan Paul, Trench, Trubner, 1929), 191; Wilkins, *Species*, 28.

3 Porphyry, *Isagoge* 3.2.10–15: "For example, about genera and species—whether they subsist, whether they actually depend on bare thoughts alone, whether if they actually subsist they are bodies or incorporeal and whether they are separable or are in perceptible items and subsist about them—these matters I shall decline to discuss, such a subject being very deep and demanding another and larger investigation." Translation by Jonathan Barnes in *Porphyry's Introduction* (Oxford: Oxford University Press, 2003), quoted in Wilkins, *Species*, 238.

> as it is, that the mind makes the pattern for sorting and naming of things, I leave it to be considered who makes the boundaries of the *sort* or *species*; since with me *species* and *sort* have no other difference than that of a Latin and English idiom.[4]

As hinted at in this quotation by Locke, there were certainly many who were busying themselves with the "sorting and naming of things" in the seventeenth and eighteenth centuries. The objects within the then popular cabinets of curiosities (or *Kunstkammern*), animal specimens, archaeological artifacts, rocks and crystals, along with exotic new finds reported in travelogues and explorers' accounts, were now to be organized, categorized, and named.[5] Scientists such as Bacon, Newton, and Descartes were enthusiastic supporters and architects of a universal language that would not only provide universal terminology for every possible thing but also classify them within ordered divisions.[6] Classifiable "things" included abstract concepts such as virtues and parts of speech, as demonstrated by John Wilkins, who includes them in an elaborate taxonomic chart published in *An Essay Towards a Real Character: and a Philosophical Language* (1668).[7]

The naturalist whom Wilkins employed to provide a taxonomy of plant and animal species for the *Essay* was John Ray, whose organization of data and definition of species according to biologically generative capacities would serve as the foundation for Linnaeus's work, so much so that one scholar states: "Without Ray's work there could have been no Linnaeus."[8] Yet while Ray may have been the first to define species biologically and to apply the

4 John Locke, *An Essay Concerning Human Understanding and Treatise on the Conduct of the Understanding* (Philadelphia: Kay, 1844), 3.5.9 (p. 286).

5 The Enlightenment Gallery in the King's Library at the British Museum is a beautiful exhibit of Enlightenment-style cabinets of curiosities, showcasing with artifacts, natural specimens, books, and tools the "seven major new disciplines of the age," which are listed as religion and ritual, trade and discovery, the birth of archaeology, art and civilization, classifying the world, ancient scripts, and the natural world. The British Museum: Explore/Galleries, Enlightenment (Room 1), accessed December 6, 2010, http://www.britishmuseum.org/explore/galleries/themes/room_1_enlightenment.aspx.

6 See Mary M. Slaughter, *Universal Languages and Scientific Taxonomy in the Seventeenth Century* (Cambridge: Cambridge University Press, 1982).

7 John Wilkins, *An Essay Towards a Real Character, and a Philosophical Language* (London: Gellibrand, 1668).

8 Charles E. Raven, *John Ray Naturalist: His Life and Works* (Cambridge: Cambridge University Press, 1942), 87.

term exclusively in a classification of living things,[9] it is Linnaeus who is remembered as the pioneer of modern taxonomy. Just as Erasmus and not the editors of the Complutensian Polyglot is usually remembered as the first to produce a printed edition of the Greek New Testament, most likely because of his edition's transformation into the tenacious *Textus Receptus*,[10] so Linnaeus is remembered because his binomial nomenclature is still in use, as well as his systematology, for better or for worse.[11] It is precisely for this reason that this history begins in detail with Linnaeus. Linnaeus's prominence in his own time marks a major shift in biological classification toward systematic and ranked taxonomy with fixed definitions of species and genus. No one of these elements is original to Linnaeus, but these, combined with the weight of his influence on all subsequent classification and naming of living beings, make him the obvious beginning for this study.

The history of New Testament textual criticism is some six hundred years shorter than that of natural classification, since its subject is considerably younger than the universe and its parts. The discussion of New Testament manuscripts and its variant readings dates perhaps as far back as the late second century CE, while the discussion of textual problems in general may be dated back to the sixth and fifth centuries BCE to Greek readers of Homer who were analyzing the texts of the *Iliad* and the *Odyssey*.[12]

The earliest evidence of biblical textual criticism comes from a pamphlet circulated during the time of Pope Victor (187–198 CE), preserved by Eusebius, which tells of a group of (perceived) heretics who "did not fear to lay hands on the divine Scriptures, alleging that they had corrected them."[13] Pride of place, however, goes to Origen of Alexandria, whose massive work

---

9 Wilkins, *Species*, 67.

10 For a brief explanation of the *Textus Receptus* and its significance, see Chapter 2.

11 See Raven, *John Ray*, 306 and Marc Ereshefsky, *The Poverty of the Linnaean Hierarchy: A Philosophical Study of Biological Taxonomy* (Cambridge: Cambridge University Press, 2001).

12 See Rudolf Pfeiffer, *History of Classical Scholarship: From the Beginnings to the End of the Hellenistic Age* (Oxford: Clarendon Press, 1968), 9–11 on Theagenes of Rhegium; 105–22 on Zenodotus; 123–51 on Callimachus. See also Leighton D. Reynolds and Nigel G. Wilson, *Scribes and Scholars: A Guide to the Transmission of Greek and Latin Literature* (Oxford: Clarendon Press, 1974), 1–15.

13 Euseb. *Hist. eccl.* 5.28.15, from *Eusebius: The Ecclesiastical History*, vol. 1, ed. and trans. Kirsopp Lake (Loeb Classical Library), 522–3: *ταῖς θείαις γραφαῖς ἀφόβως ἐπέβαλον τὰς χεῖρας, λέγοντες αὐτὰς διωρθωκέναι*. Cited and slightly modified by Bruce M. Metzger and Bart D. Ehrman, *The Text of the New Testament: Its Transmission, Corruption, and Restoration*, 4th ed. (New York: Oxford University Press, 2005), 199.

on the Old Testament, the *Hexapla*, was the product of prodigious labor and text-critical scrutiny. Although Origen seems not to have attempted the same project with the text of the New Testament, there survive many examples of his text-critical analysis of some of its variant readings.[14] Origen was well aware of the problems arising from intentional or unintentional changes made in the copying of texts:

> The differences among the manuscripts [of the Gospels] have become great, either through the negligence of some copyists or through the perverse audacity of others; they either neglect to check over what they have transcribed, or, in the process of checking, they make additions or deletions as they please.[15]

Jerome, who had his own copy of Origen's *Hexapla* and made his own corrections to it,[16] seems to have been familiar with the collation of readings, the varying worth of different manuscripts, and at least thirteen different causes for textual error.[17]

These ancient scholars were certainly looking for a correct reading of the texts they handled, a reading perhaps even equivalent to the "authentic" or "original" reading sought for in modern textual criticism. But they generally practiced an eclectic method based on what they judged, through experience and inductive reasoning, to be the right reading. No attempt was made to trace the history and genealogy of a variant, nor were manuscripts grouped into different types.

This is not so different from the methods used by the scholars who printed the first editions of the Greek New Testament. The editors of the Complutensian Polyglot and Erasmus relied upon whatever manuscripts

14 See Bruce M. Metzger, "Explicit References in the Works of Origen to Variant Readings in New Testament Manuscripts," in *Biblical and Patristic Studies: In Memory of Robert Pierce Casey*, ed. J. Neville Birdsall and Robert W. Thomson (Freiburg: Herder, 1963), 78–95.

15 Origen, *Comm. in Matt.* 15.14 (from *Origenes Werke X: Commentarius in Matthaeum I*, ed. E. Klostermann, Die Griechischen Christlichen Schriftsteller 40 [Leipzig: Hinrichs, 1935], 387–8): *πολλὴ γέγονεν ἡ τῶν ἀντιγράφων διαφορά, εἴτε ἀπὸ ῥᾳθυμίας τινῶν γραφέων, εἴτε ἀπὸ τόλμης τινῶν μοχθηρᾶς <εἴτε ἀπὸ ἀμελούντων> τῆς διορθώσεως τῶν γραφομένων, εἴτε καὶ ἀπὸ τῶν τὰ ἑαθτοῖς δοκοῦντα ἐν τῇ δορθώσει <ἢ> προστιθέντων ἢ ἀφαιρούντων*. Quoted and translated by Metzger, "Explicit References," 79–80.

16 Jer. *In Tit.* (*PL* XXVI 595; V. 734). Cited in Karl Kelchner Hulley, "Principles of Textual Criticism Known to St. Jerome," *Harvard Studies in Classical Philology* 55 (1944): 91.

17 Hulley, "Principles," 93–101.

they could access, using their best judgment to adjudicate among the variant readings.[18] Even as scholars amassed manuscripts for study and comparison, methods changed very little. John Mill's monumental edition of the Greek New Testament (1707) is an example of this;[19] though the edition shows the painstaking labor of his collations and contains insightful scrutiny of his sources, Mill does not employ methods much different from those who preceded him.[20]

Mill's edition did, however, create a new wave of panic over the stability of the sacred text as news of its rumored thirty thousand variants spread.[21] Into the fray between Free Thinkers and outraged churchmen strode the formidable Richard Bentley (1662–1742), quite assured of his own critical acumen and its absence in his opponents. In disdain of the notion that the text of the New Testament was in a precarious position, Bentley wrote:

> If there had been but One Manuscript of the *Greek* Testament, at the restoration of Learning about Two Centuries ago; then we had had no *Various Readings* at all. And would the text be in a better condition then, than now we have 30,000? So far from That; that in the best single Copy extant we should have had Hundreds of Faults, and some Omissions irreparable. Besides that the Suspicions of Fraud and Foul Play would have been increas'd immensly.[22]

In sum, more evidence is preferable to less, and, in Bentley's estimation, the number of variants posed no great danger. He predicted that upon

18 Even then, Erasmus's judgment is perhaps not so reliable, since he distrusted the best of his manuscripts. See Metzger and Ehrman, *Text of the New Testament*, 148–9.

19 John Mill, *'H KAINH ΔΙΑΘΗΚΗ Novum Testamentum* (Oxford: Sheldonian, 1707).

20 Thus Adam Fox, *John Mill and Richard Bentley: A Study of Textual Criticism of the New Testament 1675–1729* (Oxford: Blackwell, 1954), 70, concludes: "Speaking generally, the great achievement of [Mill's] Prolegomena is the choice of relevant material."

21 The first mention of there being thirty thousand variants in Mill's work seems to have been in Gerhard von Maastricht's "De Collectionibus et Collectoribus variantium Lectionum" ("On collections and collectors of variant readings") in the prolegomena of his 1711 edition of the Greek New Testament: *'H ΔΙΑΘΉΚΗ Novum Testamentum*, ed. Gerhard von Maastricht (Amsterdam: Wetsten, 1711), 25. See Fox, *John Mill and Richard Bentley*, 105, who deems this "a fairly accurate estimate."

22 Richard Bentley [Phileleutherus Lipsiensis, pseud.], *Remarks upon a Late Discourse of Free-Thinking in a Letter to F.H. D. D.*, 8th ed. (Cambridge: Bentham, [1713] 1743), 90; quoted by Fox, *John Mill and Richard Bentley*, 114.

closer inspection "there will scarce be two hundred out of so many thousands that can deserve the least consideration."[23]

Be that as it may, there still remained thousands of readings to sift through before one could determine which were significant and which were not. Collation by hand of every reading was a virtually impossible task, and the number of manuscripts, not only of the Greek New Testament but also of the Church Fathers and translations, was growing. Just as explorers were bringing a bewildering multitude of plant and animal specimens back from exotic "new" lands, so scholars amassed ancient volumes and scrolls from ransacked libraries and monasteries. The desire to analyze and make sense out of this growing body of evidence for diversity, both biological and textual, gave rise to the systematic organization of data by classification.

Classification

Locke argues that, in the face of great quantities of data, the human mind naturally organizes units of information into ranked categories:

> For the natural tendency of the mind being towards knowledge; and finding that, if it should proceed by and dwell upon only partic ular things, its progress would be very slow, and its work endless; therefore, to shorten its way to knowledge, and make each perception more comprehensive, the first thing it does, as the foundation of the easier enlarging its knowledge, either by contemplation of the things themselves that it would know, or conference with others about them, is to bind them into bundles, and rank them so into sorts, that what knowledge it gets of any of them it may thereby with assurance extend to all of that sort; and so advance by larger steps in that which is its great business, knowledge. This, as I have elsewhere shown, is the reason why we collect things under comprehensive ideas, with names annexed to them, into genera and species; i.e. into kinds and sorts.[24]

23 Richard Bentley, *Proposals for Printing a New Edition*, in *The Works of Richard Bentley, D. D.*, vol. 3, ed. Andrew Dyce (London: MacPherson [1720] 1838); quoted in *Bentleii Critica Sacra*, ed. Arthur Ayres Ellis (Cambridge: Deighton, Bell, 1862), xviii.

24 Locke, *Essay*, 2.32.6 (pp. 255–6).

Bengel, in the critical apparatus of his 1734 edition of the Greek New Testament, suggests a similar categorization into "kinds and sorts" for the task of textual criticism:

> Our collections are an abundant mass: but it is time to lay down the pen, lest the truth be exposed to the danger of ridicule. It is well that it is possible to bypass this mountain and arrive by a more level way at the means of distinguishing the codices. It is, moreover, possible to do so because of this most convenient rule: the more often not only single codices but also their lesser and greater correspondences (*syzygias*) tend toward obvious errors, the less authoritative their evidence is for difficult variants, and the more a reading is neglected by them, it should be retained as genuine.[25]

By categorizing a single codex as inaccurate through obvious error, one can categorize similar codices as faulty and thereby forgo examining each variant reading contained in each codex. This method of generalization has the advantage of speed and organization but also the corresponding disadvantage of possibly overlooking significant variants. Indeed, one of the major criticisms made by Wettstein, Bengel's rival of sorts, was that Bengel omitted far too much from his critical apparatus.[26]

Nevertheless, Bengel introduced a method of systematic categorization that is still in use today. He was the first to propose the grouping of evidence as a way to navigate and tame the "abundant mass" of New Testament texts: "The origin of the various readings could be investigated and

25 Johann Albrecht Bengel, *'H KAINH ΔIAΘHKH Novum Testamentaum Graecum* (Tübingen: Cotta, 1734), "Introductio," § 29: "Magnam conjectanea nostra silvam habent: sed manum de tabula, ne risuum periculo exponatur veritas. Bene est, quod praetergredi montem hunc, & planiore via pervenire datur ad codices discriminandos. Datur autem per hanc regulam aequissimam: Quo saepius non modo singuli codices, sed etiam syzygiae minores eorum vel majores, in aberrationes manifestas tendunt, eo levius ferunt testimonium in discrepantiis difficilioribus, eoque magis lectio ab eis deserta, tanquam genuina retineri debet." Unless otherwise noted, all translations are my own.

26 See Johann David Michaelis, *Einleitung in die göttlichen Schriften des Neuen Bundes*, vol. 1, 2nd ed. (Göttingen: Vandenhoeck, 1765), 735: "His whole collection of various readings, with the evidence in their favour, and sometime with his own sentiments upon them, he placed in his Apparatus criticus. This collection was chiefly taken from that of Mill; but Bengel omitted whatever he thought was of no importance, for which he has been censured by Wetstein, and, I believe, with reason." Translation by Herbert Marsh, from Michaelis, *Introduction to the New Testament*, 4th ed., vol. 2 (London: Rivington, 1832), 467.

shown by the individual codices, by pairs of codices, by greater and lesser groups (*syzygias*), by their families, tribes, and nations."[27] That is, readings can be identified and located within ranked orders, ranging from small ("by individual codices") to large ("nations").

It is not just of passing interest that Bengel described the largest groupings in this list in familial, ethnic, and racial terms, and that his classification is reminiscent of biological taxonomy. In the same paragraph listing "families, tribes, and nations" as orders of variants, Bengel extends the metaphor by proposing a genealogical table by which the variants and their relations can be mapped:

> And thence the relationships and discrepancies of the codices can be collected in certain tables (*schematismos*) and there can be made some concordances of these tables. And thus the whole thing can be presented to the eyes by means of a kind of quasi-genealogical chart, by means of which chart any more important or noticeable discrepancy can be, along with the crowd of its codices, found out to convince even the most reluctant doubters.[28]

Bengel's classification thereby described manuscripts as related to each other in the same way biological beings of the same species are. More specifically, he grouped texts in the same way contemporary scientists were

---

27 Bengel, "Introductio," § 29: "Posset variarum lectionum ortus, per singulos codices, per paria codicum, per syzygias minores majoresque, per famlias, tribus, nationesque illorum, investigari & repraesentari." Bengel's famous phrase, "families, tribes, and nations," is often quoted from Metzger and Ehrman, *Text of the New Testament*, 159, although there the quotation is given as "companies, families, tribes, nations." While *syzygias* might loosely be translated as "companies," it should be accompanied with "greater and lesser" as it is in the Latin. This particular phrase, "companies, families, tribes, and nations," is found as early as 1899 in Marvin R. Vincent, *A History of the Textual Criticism of the New Testament* (London: Macmillan, 1899), 88, although there it is not given as a direct quotation. The quotation in Metzger and Ehrman is furthermore erroneously cited as appearing in the "Prodromus Novi Testamenti Graeci recte cauteque ordinandi [*sic*]," as an appendix to Bengel's *Ioannis Chrysostomi de sacerdotio libri VI* (Stuttgard: Mezler and Erhard, 1725). The "Prodromus Novi Testamenti Graeci recte cauteque adornandi" is prefaced to the Chrysostom volume (it is later appended to the 1763 edition of Bengel's Greek New Testament), and it does not contain this quotation.

28 Bengel, "Introductio," § 29: "Et inde propinquitates discessionesque codicum ad schematismos quosdam reduci, & schematismorum aliquae concordantiae fieri; atque ita res tota per tabulam quandam quasi genealogicam oculis subjici, ad quam tabulam quaelibet varietas insignior cum agmine suorum codicum, ad convincendos etiam tardissimos dubitatores, exigeretur."

grouping the varieties found in the human species. The two main groups into which he divided manuscripts were the "Asiatic" and the "African":

> 1. Codices, translations, and the fathers divide into two families, the Asiatic and the African.
> 2. From the African is the Codex Alexandrinus, almost standing alone (for nearly all the African codices have been destroyed) but even so it is worth a multitude of manuscripts; with it are also included the Ethiopic, Coptic, and Latin translations. From the Asiatic are nearly all the remaining witnesses. After the Latin translations fall the Greco-Latin and Latinizing codices.[29]

Obviously, Bengel named these two text families after their geographic origins. But it is also clear that Bengel understood his classification as metaphorically *racial*. Codices sharing distinctive readings not only derive from a certain locale but Bengel further describes them as descending from "one common stock."[30] What is more, Bengel relates the geographical origin of a text with the "native color" it will have: "There are, therefore, five criteria for distinguishing a reading: the antiquity of the witnesses, their diversity, and their number: then, the origin of the false reading, and the native color of the genuine."[31]

The concept of a "common stock" and the connection between geographic location and "native color" are exactly in keeping with the discussion of human "races" in the eighteenth century. In his first edition of *Systema Naturae*, Linnaeus divides the human species into four: *Europaeus albescens, Americanus rubescens, Asiaticus fuscus,* and *Africanus*

---

29 Bengel, *'Η ΚΑΙΝΗ ΔΙΑΘΗΚΗ*, 2nd ed., ed. Philip David Burke (Tübingen: Cotta, 1763), 425: "1. Codices, Versiones, & patres in duas discedunt familias, Asiaticam & Africanam. 2. Ex Africana est cod. Al. paene solus; (quia codices Africani fere deleti sunt,) at quamlibet multis par: cum versione Aeth. Copt. Lat. Ex Asiatica ceteri fere testes. Latinae versioni subordinantur cod. Graecolatini & latinizantes."

30 Bengel, *'Η ΚΑΙΝΗ ΔΙΑΘΗΚΗ* (1734), "Introductio," § 29.3: "Codices duo pluresque, qui in libris & capitibus quibusdam singulariter congruunt . . . ab una stirpe communi descendunt, & fere pro uno codice numerari possunt" ("Two or more codices that agree remarkably in certain books and chapters descend from one common stock and can be counted virtually as one codex").

31 Johann Albrecht Bengel, *Gnomon Novi Testamenti* (Tübingen: Schamm, 1742), "Praefatio," § 8, *monita* 15: "Quinque igitur momenta sunt Decisionum: Antiquitas testium, diversitas eorum, & multitudo: tum, lectionis sequioris origo, & genuinae color nativus."

niger.[32] This division by continent and color reflects the general discussion of the human species at the time. Naturalists and philosophers alike focused on skin color, from Buffon, who stated that of human characteristics, "the first and most remarkable is colour," to Hume, who designated "negroes and in general all other species of men (for there are four or five different kinds) to be naturally inferior to the whites."[33]

As to how or why the variety of skin color came to be, naturalists had several hypotheses. Climate and other environmental influences were largely believed to affect skin color, hence the division of white, red, yellow, and black humans by continent. But differences in skin color also sparked a debate regarding whether people of such varying shades actually belonged within the same species and descended from the same creation of "Man" by God.[34] The concept of polygenism introduced the question of what "stock" or "stem" the various races derived from: did the ruddy North American, the black African, the tawny Asian, and the fair European descend not from the one, singular Adam but from different origins altogether? Buffon rejected this idea in favor of environmental factors affecting physical characteristics, pointing out that the different races are able to mix and produce fertile offspring.[35] He proposed that each species derives from a *première souche*, a "primary stock," and each species has a *moule intérieure*, an "interior mold," from which it can degenerate but to which it can also return, given the correct environmental circumstances.[36]

---

32 Carl Linnaeus, s.v. "Quadrupedia, Anthropomorpha, Homo," *Systema Naturae*, 1st ed. (Leiden: Theodore Haak, 1735).

33 Translation by J. S. Barr, *Buffon's Natural History*, vol. 4 (London: Gillet, 1807), 191. Buffon, *Histoire naturelle, générale et particulière, avec la description du Cabinet du Roy*, vol. 3 (Paris: L'Imprimerie Royale, 1749), 371: "La première & la plus remarquable de ces variétés est celle de la couleur, la seconde est celle de la forme & de la grandeur, & la troisième est celle du naturel des différens peuples." David Hume, "Of National Characters," in *Essays Moral and Political*, rev. ed. (London: Millar, 1748), excerpted in Emmanuel Chukwudi Eze, ed., *Race and the Enlightenment: A Reader* (Malden, MA: Blackwell Publishers, 1997), 33.

34 See C. Loring Brace, "One Origin or Many?" chapter 3 in *"Race" is a Four Letter Word* (New York: Oxford University Press, 2005), 37–43.

35 Buffon, "De la dégénération des animaux," in *Histoire naturelle*, vol. 14 (1766), 311: "les changemens sont devenus si grands et si sensibles, qu'il y auroit lieu de croire que le Nègre, le Lappon et le Blanc forment des espèces différentes, si d'un côté l'on n'étoit assuré qu'il n'y a eu qu'un seul Homme de créé, et de l'autre que ce Blanc, ce Lappon et ce Nègre, si dissemblans entr'eux, peuvent cependant s'unir ensemble et propager en commun la grande et unique famille de notre genre humain."

36 See Buffon, "De la dégénération des animaux."

Kant, following Buffon's lead, posited that humans belong to one "stem" or "stock" from which hybrids could be generated.[37]

These hybrids, if their physical characteristics were maintained through breeding over time, could become races, according to Kant,[38] and he proposed four races of the human species, the White, Negro, Hunnic, and Hindu races.[39] Linnaeus, as aforementioned, had divided the species into four slightly different categories, the European, American, Asiatic, and African. Blumenbach increased the number to five, and labeled them Caucasian, Mongolian, Ethiopian, American, and Malay.[40] What remains generally consistent in various divisions of the human species are the three major categories of "European," "Asiatic," and "African" races. Other categories, encompassing peoples from south and southeast Asia and the Americas were more mutable.[41] In almost all these taxonomies, the White, European, or Caucasian race was presumed superior or primary, while others were inferior, derivative, or both.[42]

New Testament textual critics would likewise classify and reclassify codices, and in this Bengel's metaphor of texts as belonging to "families, tribes, and nations" would prove fundamental. Bengel's classification

---

37 Immanuel Kant, "Von den verschiedenen Racen der Menschen," in *Immanuel Kant: Hauptwerke*, vol. 2, *Vorkritische Schriften 1757–1777*, ed. Königlich-Preussische Akademie der Wissenschaften zu Berlin (Charlottesville, VA: InteLex, [1775] 1999), 430: "Auf diese Weise sind Neger und Weiße zwar nicht verschiedene Arten von Menschen (denn sie gehören vermuthlich zu einem Stamme), aber doch zwei verschiedene Racen: weil jede derselben sich in allen Landstrichen perpetuirt, und beide mit einander nothwendig halbschlächtige Kinder oder Blendinge (Mulatten) erzeugen."

38 Ibid., 429: "Unter den Abartungen, d.i. den erblichen Verschiedenheiten der Theire, die zu einem einzigen Stamme gehören, heißen diejenigen, welche sich sowohl bei allen Verpflanzungen (Versetzungen in andre Landstriche) in langen Zeugungen unter sich beständig erhalten, als auch in der Vermischung mit andern Abartungen desselbigen Stamms jederzeit halbschlächtige Junge zeugen, Racen."

39 Ibid., 432.

40 Johann Friedrich Blumenbach, *De generis humani varietate nativa*, 3rd ed. (Göttingen: Vandenhoek and Ruprecht, 1795), esp. 285–7.

41 Parallels in racial and ethnic categories exist today. The US Census Bureau includes five racial categories: White, Black or African American, American Indian or Alaska Native, Asian, and Native Hawaiian or Other Pacific Islander. See "Race: About this Topic," United States Census Bureau, accessed July 19, 2015, http://www.census.gov/topics/population/race/about.html. According to the site, however, these definitions are not meant to imply biological categories but social ones. Peoples who do not self-identify as any of these categories, however, are still relegated to a type of miscellany.

42 See Chapter 2.

was ultimately bipartite, dividing manuscripts and versions into a more ancient family, the African, and a later family, the Asiatic.[43] Semler at first took over Bengel's two "nations," calling the African "Western" or "Egyptian-Palestinian," and the Asiatic "Eastern."[44] He later decided on three classes, the earlier Alexandrian (Syriac, Coptic, and Ethiopic versions) and Western (Latin versions) classes, and the later Eastern (Antiochian and Constantinopolitan).[45] Semler's student, Griesbach, settled upon those three divisions as well (two ancient and the third later), after having divided them into as many as six groups.[46] This division of texts into earlier and later families continued through the work of Hug, Scholz, Westcott and Hort, and beyond, with variations on a theme introduced not only because of differing conjectures but also because of a growing body of evidence.

As these classifications developed, they represented geographic origins less than textual similarities. Nevertheless, it is clear that the geographic names continued to evoke lands and regions together with their imagined peoples and cultures. In 1831 Lachmann published his groundbreaking edition of the Greek New Testament, the first to leave behind the *Textus Receptus* altogether. Lachmann's incredibly brief explanation of his sources at the end of the text indicated the manuscripts of "the oldest

43 See Samuel Prideaux Tregelles, *An Account of the Printed Text of the Greek New Testament* (London: Bagster, 1854), 71. For a highly useful summary of Bengel's bipartite (versus his earlier tripartite) classification and its subsequent influence, see Eldon J. Epp, "Textual Clusters: Their Past and Future in New Testament Textual Criticism," in *The Text of the New Testament in Contemporary Research: Essays on the Status Quaestionis*, ed. Bart D. Ehrman and Michael W. Holmes, 2nd ed., New Testament Tools, Studies and Documents 42 (Leiden: Brill, 2013), 523–34.

44 Johannes Salomo Semler, "Vorrede," in *Hermeneutisches Vorbereitung*, vol. 3, part 1 (Halle: Hemmerde, 1765): "Ausser diesen Zeugnissen, habe ich aus der Vergleichung der Handschriften ebenfals mehr leicht gesammelt für diese Sache; und habe besonders sehr deutlich und fast unwidersprechlich gezeiget: daß es unter den Handschriften, die wir noch übrig haben, sehr merkliche Spuren davon gebe, daß ehedem es wenigstens zweyerley Recension des griechischen Textes gegeben habe; eine kan man die morgenländische oder Luciani, nennen, (blos um kurz zu reden;) die andere die abendländische, egyptische, palästinische, Origenianische. Diese ist gemeiniglich kürzer und mit andern Redensarten abgefasset, als jene; sie ist ganz und gar einstimmig mit der ältesten latina, oder Itala, oder welche da war, ehe Vulgata zu herrschen anfieng."

45 Semler, *Apparatus ad liberalem Novi Testamenti* (Halle: Tramp, 1767), esp. §20. See also Metzger and Ehrman, *Text of the New Testament*, 161–2.

46 Metzger and Ehrman, *Text of the New Testament*, 165. See also Epp, "Textual Clusters," 530 on Griesbach's ultimate tripartite (and not bipartite, contra Tregelles) scheme of textual families.

eastern churches" as those he consistently followed, although he also consulted those of "Italy and Africa."[47] Scholars who had not read the fuller explanation of his method (and these were many since it had been published a year earlier exclusively in a German journal)[48] assumed that by "eastern" (*oriens*) Lachmann meant Bengel's *less* ancient "Asiatic" nation, Semler's "Eastern," Griesbach's "Constantinopolitan," and so on.[49]

As a matter of fact, Lachmann meant nothing of the sort: he was referring to *early* ancient codices, including Codex Alexandrinus, Codex Vaticanus, and Codex Ephraemi Rescriptus.[50] Regardless of the provenance of these codices (which remains uncertain), the term "eastern" or *oriens* evoked then, as they do now, a set of geographies, aesthetics, and cultures in the European imagination. In its broadest sense, the East or Orient represents the boundaries of European knowledge, exploration, and colonization—the point of contrast for the "West."[51] In keeping with this oppositional binary is the notion that Christendom, and its history, can be divided into East (Orthodox/Byzantine) and West (Catholic/Protestant), with European identification always centering around the latter.[52] The confusion over Lachmann's use of "eastern" demonstrates that these labels and classifications did not signal simple, geographical regions but unstable, imagined geographies signifying imagined races and cultures.

This is made even clearer in Hort's descriptions of different text groups, the Western, Alexandrian, Neutral, and Syrian, in his introduction to

---

47 Karl Lachmann, *Novum Testamentum Graece*, vol. 1 (Berlin: Reimer, 1831), 491: "hic satis erit dixisse, editorem nusquam iudicium suum, sed consuetudinem antiquissimarum Orientis ecclesiarium secutum esse. hanc quoties minus constantem fuisse animadvertit, quantum fieri potuit ea quae Italorum et Afrorum consensu comprobarentur praetulit: ubi pervagatam omnium auctorum discrepantiam deprehendit, partim uncis partim in marginibus indicavit."

48 Lachmann, "Rechenschaft über seine Ausgabe des Neuen Testaments," *Theologische Studien und Kritiken* 3 (1830): 817–45.

49 Tregelles, *Account*, 97–9.

50 Lachmann, "Rechenschaft." For a comprehensive list, see Epp, "Textual Clusters," 535.

51 See Edward Said, *Orientalism* (New York: Vintage Books, 1978) and Martin W. Lewis and Kären E. Wigen, *The Myth of Continents: A Critique of Metageography* (Berkeley: University of California Press, 1997), 47–103.

52 This may seem like common sense, but consider that Greece is a contested geographic area in terms of cultural and religious identification in European (an unstable term itself) scholarship. Classical Greece is co-opted by German scholarship, for example, as part of the "West's" intellectual genealogy, whereas post-classical Greece and Greek Orthodoxy is decidedly viewed as Other. See Anthony Smith, *The Ethnic Origin of Nations* (Oxford: Blackwell, 1986), 203, and also Chapter 3 on philhellenism in biblical studies.

his and Westcott's authoritative edition of the Greek New Testament in 1882.[53] He sets forth the characteristics of each group not only in terms of grammar and literary features but also in adjectives strikingly similar to race descriptions of the eighteenth century. The Western text displays "a certain rapid vigour and fluency" in contrast to the "calm and reticent strength of apostolic speech."[54] The Alexandrian text, on the other hand, is the product of "careful and leisurely hands."[55]

Hort's characterization of the Syrian text is especially anthropomorphic and characteristic of the Orientalism of his time:

> As distinguished from the bold vigour of the "Western" scribes, and the refined scholarship of the Alexandrians, the spirit of its own corrections is at once sensible and feeble. Entirely blameless on either literary or religious grounds as regards vulgarised or unworthy diction, yet shewing no marks of either critical or spiritual insight, it presents the New Testament in a form smooth and attractive, but appreciably impoverished in sense and force, more fitted for cursory perusal or recitation than for repeated and diligent study.[56]

Here, portrayed in contrast to the dynamism and acuity of the Western and Alexandrian texts, Hort presents the Syrian text in stereotypically feminine terms: it is attuned to emotions and the sensual ("sensible")[57] but weak ("feeble"). The text is imagined almost explicitly as a female character, who is above coarseness, docile, and without much depth, but presenting an erotic physical form, "smooth and attractive."

---

53 Brooke Foss Westcott and Fenton John Anthony Hort, ed., "Introduction," in *The New Testament in the Original Greek* (Cambridge: Macmillan, 1882). In discussing concepts explained at length in the Introduction, I reference only Hort, since he is its primary author. For broader principles upon which their edition of the NT is based, such as the supremacy of the "neutral" text, I reference both Westcott and Hort.

54 Ibid., 122.

55 Ibid., 131. Compare with the strictly statistical approach of James R. Royse, *Scribal Habits in Early Greek Papyri*, New Testament Tools, Studies and Documents 36 (Leiden: Brill, 2008).

56 Ibid., 135.

57 It is clear here that Hort means "sensible" in this older sense, now out of use, rather than the current popular meaning of "sensible" as practical and using good judgment, given the context.

This "eastern" text stretches languidly before the philologist, whose gaze appreciates its aesthetic beauty but deems its substance anemic and lacking.[58] This precise dynamic of "Eastern" object (or, indeed, of any "Other" object of study) and "Western" scholar is crystallized in this statement by historian Edgar Quinet in 1842: "Asia has its prophets; Europe its doctors."[59] On one side lies the exotic, mystical, even *Byzantine* object, on the other stands the analytical, rational, historical subject.[60]

The gaze of the European scholar, analyzing both texts and peoples from above, made clinical observations not only of supposedly quantifiable distinctions (such as the addition of pronouns in a text or the hair and skin color of humans) but assigned to them dispositions, personalities, and moral and intellectual characteristics. Hort's descriptions of text types echoes earlier catalogues of racial and national features. Linnaeus, for example, displays the same Orientalism by describing the European race as "gentle, acute, innovative . . . governed by rites," and the Asiatic race as "harsh, haughty, covetous . . . governed by opinions."[61] Kant likewise makes such pronouncements:

> The Japanese could in a way be regarded as the Englishmen of this part of the world, but hardly in any other quality than in their resoluteness—which degenerates into the utmost stubbornness—their valor, and disdain of death. For the rest, they display few signs of a finer feeling. The Indians have a dominating taste of the

58 See Said, *Orientalism*, 137, on both Edgar Quinet and Ernest Renan: "As a scholarly attitude the picture of a learned Westerner surveying as from a peculiarly suited vantage point the passive, seminal, feminine, even silent and supine East, then going on to *articulate* the East, making the Orient deliver up its secrets under the learned authority of a philologist whose power derives from the ability to unlock secret, esoteric languages—this would persist in Renan."

59 Edgar Quinet, *Du Génie des Religions* (Paris: Charpentier, 1842), 53: "L'Asie a les prophètes; l'Europe a les docteurs."

60 This understanding of East/West can be found explicitly stated in scholarship even more than a century after Quinet. See Filmer Stuart Cuckow Northrop, *The Meeting of East and West* (New York: Macmillan, 1960): "the meaning of Eastern civilization in its relation to the meaning of Western civilization is as follows: *The Orient, for the most part, has investigated things in their aesthetic component; the Occident has investigated these things in their theoretic component.*" Emphasis original. Quoted in Lewis and Wigen, *The Myth of Continents*, 77.

61 Linnaeus, *Systema Naturae*, 10th ed. (Halle: Kurt, 1760), 20: "Europaeus. Levis, acutissimus, inventor. Regitur Ritibus. Asiaticus. Severus, fastuosus, avarus. Regitur Opinionibus."

> grotesque, of the sort that falls into the adventurous. Their religion consists of grotesqueries. . . . What trifling grotesqueries do the verbose and studied compliments of the Chinese contain![62]

Here, in both Linnaeus and Kant, the Eastern peoples (including the Far East in Kant) are again imagined as the same shallow, fanciful, aesthetically driven, and mercurial race, serving as a foil for the West. The fact that these are imagined and projected traits, without even the assistance of direct observation and experience,[63] is made especially clear when one considers that Kant refused to travel more than a few miles from his birthplace, Königsberg, the span of his entire eighty years.[64] Yet he and other scholars assumed unshakeable confidence that they could not only quantify distinctive physical features of vast groups of people but also determine specific dispositions represented by them, presumably all from travelogues and the occasional "specimen."[65]

The physical traits that were supposedly the basis for these racial and national characteristics were often listed in exhaustive detail, with particular focus on facial features (physiognomy) and skin color. For example,

62 Translation by John T. Goldthwait from Immanuel Kant, *Observations on the Feeling of the Beautiful and Sublime* (Berkeley: University of California Press, 1960), excerpted in Eze, *Race and the Enlightenment*, 55. Kant, "Beobachten Über das Gefühl des Schönen und Erhabenen," in *Hauptwerke*, 2:252: "Die Japoneser könnten gleichsam als die Engländer dieses Welttheils angesehen werden, aber kaum in einer andern Eigenschaft, als ihrer Standhaftigkeit, die bis zur äußersten Halsstarrigkeit ausartet, ihrer Tapferkeit und Verachtung des Todes. Übrigens zeigen sie wenig Merkmale eines feineren Gefühls an sich. Die Indianer haben einen herrschenden Geschmack von Fratzen von derjenigen Art, die ins Abenteuerliche einschlägt. Ihre Religion besteht aus Fratzen . . . Welche läppische Fratzen enthalten nicht die weitschichtige und ausstudirte Complimente der Chineser."

63 These would, of course, also be dubious sources of information, given that the entire project of cataloguing "races" already assumes ontological and hierarchical differences between people groups.

64 My thanks to Michael J. Thate for bringing this disparity between Kant's experience and presumption to my attention.

65 The most famous "specimen" was Saartjie (Sara) Baartman, a Khoisan woman from South Africa, who was put on exhibition as the "Hottentot Venus" in both England and France. The French naturalist Georges Cuvier examined her while she was in Paris, and, after her death, dissected her and displayed her skeleton and genitals (cast in wax) in the Muséum d'Histoire Naturelle and then the Musée de l'Homme until 1974. See Sadiah Qureshi, "Displaying Sara Baartman, The 'Hottentot Venus,'" *History of Science* 42 (2004): 233–57; and Clifton Crais and Pamela Scully, *Sara Baartman and the Hottentot Venus: A Ghost Story and a Biography* (Princeton: Princeton University Press, 2009).

Buffon describes the Laplanders, Borandians, Samoiedes, Ostiacks, and Greenlanders with great attention to facial detail:

> These people have broad faces and flat noses; their eyes are of a yellowish brown, inclining to black, their eye-lids are drawn toward the temples, their cheek-bones are extremely prominent, their mouths are large, the lower part of their countenance narrow, their lips thick and turned outward; their voices are shrill, with heads bulky, hair black and straight, and skin of a tawny colour.[66]

He concludes that all these nations belong to one race and evaluates their social and moral character: "Incivility, superstition, and ignorance, are alike conspicuous in them all."[67]

Hort himself uses this type of description in writing his opinion of the plight of slaves in America during the American Civil War:

> The refusal of education to them is abominable; how far they are capable of being ennobled by it is not so clear. As yet *everywhere* (not in slavery only) they have surely shown themselves only as an immeasureably inferior race, just human and no more, their religion frothy and sensuous, their highest virtues those of a good Newfoundland dog. If enjoyment and comparative freedom from sorrow and care make up happiness, probably no set of men in Europe (unless it be the Irish) are so happy.[68]

Again the Western scholar confidently surveys an imagined category and pronounces totalizing ("*everywhere*") judgments that invariably veer toward descriptions of the sensuous, placid, and fatuous. This pattern of hierarchical classification followed by condescending dissection is the discursive trope used by Hort on text types, incorporated from the racial taxonomy of his time.

---

66 Translation by Barr, *Buffon's Natural History*, 4: 192. Buffon, *Histoire naturelle*, 3:400–1: "Tous ces peuples ont le visage large & plat, le nez camus & écrasé, l'iris de l'œil jaune-brun & tirant sur le noir, les paupières retirées vers les temples, les joues extrêmement élevées, la bouche très-grande, le bas du visage étroit, les lèvres grosses & relvées, la voix grêle, ta tête grosse, les cheveux noirs & lisses, la peau basanée."

67 Translation by Barr, *Buffon's Natural History*, 4: 194. Buffon, *Histoire Naturelle*, 3: 402: "ils sont tous également grossiers, superstitieux, stupides."

68 Fenton John Anthony Hort, *Life and Letters of Fenton John Anthony Hort*, ed. Arthur Fenton Hort (London: Macmillan, 1896), 458, in a letter to the Rev. John Ellerton, September 25, 1862. Emphasis original.

Conclusion

It is clear, then, that Bengel's racial metaphor set the tone and structure for manuscript grouping and categorization, and this legacy is still evident in the language of current textual critics. Terms such as the "physiognomy" of a text, "textual consanguinity," and "textual complexion" are quite common in grouping and describing textual characteristics.[69] From these examples it is obvious that textual critics have not only borrowed the structure of racial classification but in borrowing it transformed the texts of their study into racialized, biological beings.

These textual beings are fully incarnate, with skin color and facial features, active bodies or languorous ones. Perhaps most important to the philologist, however, is the understanding that these bodies can be harmed and broken, and that the discovery of such damage to the flesh is the task of the textual critic. Lachmann in his second edition of the Greek New Testament (1850), for example, refers to a difficult reading as a "wound," and the smoothed-over reading in later versions and the *Textus Receptus* as "scars and bandages" that mask the original damage.[70] The untouched wound, Lachmann implies, is preferable to a patching-up of flesh because the skilled scholar, now serving as physician, can better attend to an untreated cut than a badly healed scar.

So Tregelles summarizes the argument in his survey of the field in 1854, further explaining the metaphor of the wounded textual body:

> Lachmann speaks of those who prefer to see the text "skinned over and plaistered," rather than with the wounds visible: that is, that some would prefer the text as it has passed through the hands of copyists and non-critical editors, with the wounds (if such there be) of the

---

69 See, for example, Bart D. Ehrman, "Methodological Developments in the Analysis and Classification of New Testament Documentary Evidence" in *Studies in the Textual Criticism of the New Testament*, New Testament Tools and Studies 33 (Leiden: Brill, 2006), 9, 15; Metzger and Ehrman, *Text of the New Testament*, 184, 185; Royse, *Scribal Habits*, 16.

70 Lachmann, *Novum Testamentum Graece et Latine*, vol. 2 (Berlin: Reimer, 1850), VIII–IX: "Actuum 13,32 *καὶ ἡμεῖς ὑμᾶς εὐαγγελιζόμεθα τὴν πρὸς τοὺς πατέρας ἐπαγγελίαν γενομένην, ὅτε ταύτην ὁ θεὸς ἐκπεπλήρωκεν τοῖς τέκνοις ΗΜΩΝ, ἀναστήσας Ἰησοῦν, ὡς καὶ ἐν τῷ ψαλμῷ γέγραπται τῷ πρώτῳΥἱός μου εἶ σύ, ἐγὼ σήμερον γεγέννηκά σε.* summus testium consensus non efficit ut illud *ἡμῶν* verum sit, sed ut proximum vero. hoc qui aut omittunt aut cum libris vulgaribus faciunt *αὐτῶν ἡμῖν*, cicatrices et linamenta malunt quam vulnera. verum hoc esse existimo, *ΕΦ' ΗΜΩΝ ἀναστήσας Ἰησοῦν*, 'nostro tempore.' nunc, inquit, illud certum tempus atque ille hodiernus dies adest, quo deus se filium genuisse dixit."

> earliest copies and versions concealed by a sort of artificial vail. . . . [The common reading] seems to have only sprung up as an amendment, a "skinning over and bandaging". . . .[71]

Here, Tregelles specifically uses medical terminology, "skinning over," rather than simply, "scars" (*cicatrices*) in the translation of Lachmann's text, referring to the production of new skin by the body itself in the process of healing.[72] A close reading of Tregelle's description might identify the "skinning over" with the work of copyists, whose repeated by-rote copying can introduce changes slowly and be compared to the natural, incremental self-healing of the "skin" of the text. The "plaister" and "bandaging," on the other hand, is the work of "non-critical editors" who intentionally place amendments in the text in order to improve it with "an artificial vail." In either case, the text itself is a wounded physical body in need of healing.

Beyond flesh wounds, the embodied text is also imagined to suffer from other ailments that the textual critic can treat. The classical philologist A. E. Housman describes the variety of text-critical methods as "remedies":

> We must have no favourite method. An emendator with one method is as foolish as a doctor with one drug. The scribes knew and cared no more about us and our tastes than the diseases care about the tastes of doctors; they made mistakes not of one sort but of all sorts, and the remedies must be of all sorts too.[73]

With Housman the metaphor expands so that scribes are cast as analogous to diseases, wreaking havoc on the texts they handle. And the analogy of the ailing text incarnate continues in contemporary text-critical

---

71 Tregelles, *Account*, 111–2.

72 See John Hunter, "Of Skinning," in *A Treatise on the Blood, Inflammation, and Gunshot Wounds* (Philadelphia: Haswell, Barrington, and Haswell, [1794] 1840), 470–4. My thanks to Paul Craddock for pointing me to Hunter's work and clarifying the difference between "skinning over" (production of skin by the body's own healing mechanism) and active skin grafting.

73 A. E. Housman, ed., *M. Manilii Astronomicon liber primus* (Cambridge: Cambridge University Press, 1903), liii. Quoted in Metzger and Ehrman, *Text of the New Testament*, 305.

scholarship. Metzger begins the chapter on textual error, in his classic work *The Text of the New Testament*, using this exact metaphor:

> As the physician must make a correct diagnosis of a disease before attempting to effect its cure, so the textual critic must be aware of the several kinds of injury and danger to which a text transmitted by handwriting is liable to be exposed before trying to rectify the errors.[74]

The text as a racialized and literally embodied being is thus, like any biological being, subject to "injury and danger." Beyond suffering bodily harm at the hands of scribes and editors, moreover, it, as a supine, feminine figure, is subject to the further dangers of corruption and contamination through sexual penetration and adulteration. The "unfaithful" scribe can introduce foreign materials or readings from an entirely different text group, creating contaminated, even bastard offspring.[75]

These concepts of purity under threat touch directly upon the idea of the racialized body, especially in eighteenth- and nineteenth-century Europe, where ("Caucasian") racial purity was perceived to be under threat through the mixing of blood lines. The possible interrelationship of the different races and text groups, their genealogies and interbreeding, and the search for a mythical, pure, original archetype were the natural result of classification by race.

---

74 Metzger and Ehrman, *Text of the New Testament*, 250.

75 In this increasingly complex analogy, scribes can be seen as both the adulterating agent who creates new, bastard offspring or the unscrupulous matchmaker who creates a new reading by combining two different readings from two different text groups. The accuracy of scribes is most often described in terms of "faithfulness," a term whose range of meaning includes sexual, marital, and reproductive exclusivity and devotion. For examples of scribal "faithfulness," see Bernhard Blumenkranz, "Fidélité du scribe," *Revue du Moyen Âge* 8 (1952): 323–6. For the social contexts of scribes and the conditions of their work, see Kim Haines-Eitzen, *Guardians of Letters: Literacy, Power, and the Transmitters of Early Christian Literature* (Oxford: Oxford University Press, 2000).

2

Lachmann and the Genealogy and Corruption of Texts

It does no good to say that Abraham will have more descendants than Isaac or Jacob, not to mention David or Hezekiah. The question is whether the majority of Abraham's descendants, through either Isaac or Esau, continued through the years to look more like their original ancestor than like their more immediate ancestors through the phenomenon of mixed parentage.

GORDON D. FEE, "The Majority Text and the Original Text of the New Testament"

WHAT DO THE Jewish patriarchs have to do with textual criticism? And how is their family resemblance of any significance? The quotation above comes from Gordon Fee's incisive demolition of the Majority Text as the closest to an "ancestral" text of the Greek New Testament and therefore the most "trustworthy."[1] Its name comes from the fact that the vast majority of extant manuscripts preserve its text type (which Westcott and Hort called the Syrian, and others the Byzantine or Standard Text), and it comprises a group of texts that became dominant in the Byzantine Empire around the seventh century.[2]

The impetus for defending the Majority Text derives from the belief that the *Textus Receptus* (and its English translation, the King James Version) most truly represents the word of God. The *Textus Receptus* should

1 The epigraph comes from Gordon D. Fee, "The Majority Text and the Original Text of the New Testament," in *Studies in the Theory and Method of New Testament Textual Criticism*, ed. Eldon J. Epp and Gorden D. Fee, Studies and Documents 45 (Grand Rapids, MI: Eerdmans, 1993), 185. Ibid., 183–208.

2 See Bruce M. Metzger and Bart D. Ehrman, *The Text of the New Testament: Its Transmission, Corruption, and Restoration*, 4th ed. (New York: Oxford University Press, 2005), 279–80.

not be conflated with the Majority Text. The latter is a group of texts classified together because of shared readings, while the *Textus Receptus* is an edition of the Greek New Testament *based mostly on* the Majority Text. The story of how this edition came to be revered as "the received text," as if received from the very hand of God, is a curious one that can only be related here briefly.[3]

From what historians can discern, Dutch scholar Desiderius Erasmus, with the support of publisher Johann Froben in Basel, took on the task of producing an edition of the Greek New Testament sometime between 1512 and 1514.[4] Production proceeded at a furious pace so that when the entire volume was available in March 1516, it beat out the Complutensian Polyglot's New Testament (already printed in 1514 but not authorized for circulation until 1520)[5] as the first edition published and available on the market. Because of the rush to print, the scarcity of older manuscripts at Basel, and perhaps Erasmus's own shortcomings as a critical philologist,[6] the first edition was full of typographical errors and based upon a handful of manuscripts whose readings (by modern text critical standards) have little authority. Perhaps most emblematic of the conjunction of all these elements are the last six verses of the Apocalypse in Erasmus's edition. The twelfth-century manuscript he used for the text lacked these last few verses and, taking liberties no editor would dream of today, Erasmus simply referred to the verses in the Latin Vulgate and back-translated them into Greek.[7]

Nevertheless, thousands of copies were printed (both authorized and unauthorized) and the text that Erasmus produced (although he later improved the Greek somewhat in the fourth edition after consulting the Complutensian Polyglot)[8] became the basis for further editions all throughout Europe for the next four hundred years. An edition, printed in 1633 by publishers

3 For a more detailed narrative, see ibid., 137–64.

4 Ibid., 142–3, who state that Erasmus discussed the edition in 1514, "probably not for the first time," with Froben, after which there was a break in correspondence before negotiations began again in spring of 1515. Compare this to Leighton D. Reynolds and Nigel G. Wilson, *Scribes and Scholars: A Guide to the Transmission of Greek and Latin Literature*, 3rd ed. (Oxford: Clarendon Press, 191), 160, who state that Erasmus began work on the edition, with four Greek manuscripts at hand, during 1512–13 in England.

5 Metzger and Ehrman, *Text of the New Testament*, 139.

6 See Reynolds and Wilson, *Scribes and Scholars*, 160–1.

7 Metzger and Ehrman, *Text of the New Testament*, 145.

8 Ibid. 148–9.

Bonaventura Elzevir and Abraham Elzevir, carried a note of advertisement stating that "[the reader has] the text now received by all, in which we give nothing changed or corrupted."[9] This designation of the "received text" took on a sacred significance the advertisers had no intention of implying, and the *Textus Receptus* became revered as tantamount to the very words breathed by God. By extension, the Majority Text upon which it is based also came to be understood as the most authoritative and trustworthy text group. When scholars such as Lachmann and Westcott and Hort rejected the Majority Text as inferior to other, more ancient texts (as discussed below), their work was decried not (only) as poor scholarship but as sacrilege.[10]

The defense of the Majority Text (and, implicitly, the *Textus Receptus*) continues to the present, and this is what Fee counters in the epigraph at the beginning of this chapter. The specific argument he addresses posits that the Majority Text is the text type closest to the "original text" of the New Testament because its extant readings are by far the most numerous. The scholar behind the argument, Zane Hodges, arrives at this conclusion by assuming a principle using the metaphor of family genealogy, namely that "the copies nearest the autograph will have the largest number of descendants."[11]

The problem with this principle, as Fee points out, is that it is simply not true—not even on the level of the metaphor since a direct offspring of the ancestor (Abraham, in this case) could be childless, but that would not make the offspring less related to the ancestor than another who spawned a multitude. In other words, the rate of reproduction is not a constant for human families, and even less so for texts and manuscripts through centuries of hand copying. Furthermore, and this is Fee's main point, a foremost canon of textual criticism is that readings should be weighed, not counted,[12] that is, the quality of a reading is far more important than

9 Quoted ibid., 152: "Textum ergo habes, nunc ab omnibus receptum: in quo nihil immutatum aut corruptum damus."

10 See, for example, John William Burgon, *The Revision Revised* (London: Clowes, 1883).

11 Zane C. Hodges, "The Greek Text of the King James Version," in *Which Bible?* ed. David Otis Fuller (Grand Rapids, MI: Grand Rapids International Publications, 1970), 21, quoted by Fee, "The Majority Text," 184.

12 Derived from Bengel's proposition that texts be classified in order to deal more easily with masses of readings (see Chapter 1) but Politian, before Bengel, also proposed that "the testimonies of the ancients should not so much be counted up as weighed." Quoted in Anthony Grafton, *Joseph Scaliger: A Study in the History of Classical Scholarship*, vol. 1 (Oxford: Oxford University Press, 1983), 26.

how many of that reading exists. Even if one text were copied a thousand times over, that is, if it gave rise to thousands of "descendants," if its readings are of little value, its descendants, no matter how numerous, lend it no authority.

Both Fee and Hodges, however, share the same assumption: that the value and authority of any reading derives from its proximity to an ancestral autograph. Furthermore, both argument and counterargument make use of the metaphors of family descent, genealogy, mixed parentage, and family resemblance with ease. In fact, Fee gives an extended analogy using jumbled constructions of nationality, race, ethnicity, and geography to illustrate what probably happened in the transmission of the Majority Text:

> But the fact is that in both human and manuscript lineage the further away one gets from the original "parent(s)" the more the lineage picks up cross-fertilization, especially when one throws in the clear geographic changes brought about by easy mobility.
>
> Thus in response . . . I suggested the following analogy: Let us assume that two very Swedish parents (tall, blond, blue eyes, and so forth), Olaf and Helga Olson, have two boys, Karl and Sven. These two sons also marry Swedish young women, and each has two children. One of Karl's sons emigrates to the United States and marries a very Italian young woman (short, black hair, dark eyes, and so on), and this son happens to be prolific and has twelve children. Of these children seven marry Italians, one marries a Swede, and four marry American "mixtures." Meanwhile, in Sweden, catastrophe has struck the house of Olson and the two children of Sven are killed in an accident. The other son of Karl had one child, a daughter, who turned out to be barren. Thus the only descendants of Olaf Olson are in America, and the vast majority are now assuming Italian features.
>
> That something very much like this happened to the NT manuscript tradition is precisely why the argument from numbers is so totally irrelevant.[13]

Here, as in other examples from textual criticism discussed in the previous chapter, text types are analogous to racial types characterized by

13 Fee, "The Majority Text," 185 n. 6.

stereotypical profiles. In this analogy the ancestral family is "very Swedish . . . (tall, blond, blue eyes, and so forth)" while the numerous offspring that bear little resemblance to the "original" family are "very Italian . . . (short, black hair, dark eyes, and so on)."[14]

The most significant aspect of Fee's analogy is not the different ethnic/textual types, however, but their ability to intermingle and to reproduce. This reproductive "mixture" between text types separated by taxonomic classification is the leading metaphor used by textual critics to describe the history of textual transmission. For example, one textual critic writes that the development of distinct textual characteristics in a region "might be explained by a biological analogy: selective breeding enhances particular characteristics."[15] As is clear from the last quotation, texts have not only been racialized but they have also been sexualized. They are able to "breed," generating different families of offspring as they mix their racial lines.

Genealogy and Corruption

The description of relationships between texts and manuscripts as *familial* is certainly as old as Bengel—the classification of readings was not Bengel's only innovation. As mentioned in the previous chapter, he also proposed the drawing up of genealogical tables by which the relationships between codices and their differences could be articulated:

> And thence the relationships and discrepancies of the codices can be collected in certain tables (*schematismos*) and there can be made some concordances of these tables. And thus the whole thing can be presented to the eyes by means of a kind of quasi-genealogical chart, by means of which chart any more important or noticeable

14 Fee's choice of words and ethnic stereotypes offer several tantalizing avenues for discussion in regards to notions of race/ethnicity. For example, the use of "very" to describe someone as "very Swedish" or "very Italian" assumes a quintessential physical profile that is nationalistic. This leads one to wonder what Fee means with regards to "American 'mixtures'" and whether he has a national-ethnic profile of "America" in mind. Additionally, it should be noted that the ancestral family is fair in coloring, whereas mixture introduces darker coloring so that the "original" family characteristics are obscured. The choice of ancestral coloring may be merely coincidental, but given the history of concepts such as racial degeneration and corruption as discussed in this chapter, the coincidence is telling.

15 David C. Parker, *The Living Text of the Gospels* (Cambridge: Cambridge University Press, 1997), 23.

> discrepancy can be, along with the crowd of its codices, found out to convince even the most reluctant doubters.[16]

This proposal indicated a giant stride away from a heavy reliance on *emendatio*, which is the correction of texts relying upon the conjecture of the editor, who may or may not consult other manuscripts to adjudicate which reading is preferable.[17] What Bengel was proposing suggested the use of *recensio*, the determination of which texts provide the best reading by attempting to take into account the history of textual transmission.[18] The use of "quasi-genealogical charts" was indicative of Bengel's conviction that it was not the number of witnesses for a reading that lent it weight but the trustworthiness of the family or group from which that reading came.

The understanding of the importance of "genealogical" ties or "kinship" between texts dates back at least to the Italian Humanist Politian (1454–1494).[19] Politian certainly recognized the significance of an older manuscript from which other manuscripts were derived, that the errors and readings of the older manuscript would be carried by its copies and, no matter how numerous its copies were, they would not present a different tradition:

> I have come upon a text of the *Silvae* of Statius . . . from which alone all the other available manuscripts seem to have emanated, though it is full of errors and corrupt and, what is more, as I think, reduced to half its original extent.[20]

---

16 Johann Albrecht Bengel, *'H KAINH ΔIAΘHKH Novum Testamentum Graecum* (Tübingen: Cotta, 1734), "Introductio," § 29: "Et inde propinquitates discessionesque codicum ad schematismos quosdam reduci, & schematismorum aliquae concordantiae fieri; atque ita res tota per tabulam quandam quasi genealogicam oculis subjici, ad quam tabulam quaelibet varietas insignior cum agmine suorum codicum, ad convincendos etiam tardissimos dubitatores, exigeretur."

17 See Sebastiano Timpanaro, *The Genesis of Lachmann's Method*, ed. and trans. Glenn W. Most (Chicago: University of Chicago Press, 2005), 46, which describe two types of *emendatio*: "*emendatio ope codicum* [emendation with the help of manuscripts] and *emendatio ope ingenii* [emendation with the help of native wit] or *coniecturae* [of conjecture]." See also n. 3 on the same page.

18 According to Metzger and Ehrman, *Text of the New Testament*, 161 n. 58, Semler was the first to use the term *recension* to indicate a group of texts.

19 Timpanaro, *Genesis*, 47–51.

20 Quoted in Timpanaro, *Genesis*, 47, as the *subscriptio* of the incunable of Statius's *Silvae*. See also Reynolds and Wilson, *Scribes and Scholars*, 144.

Despite producing perhaps the most famous edition based on *emendatio*—the *Textus Receptus*—Erasmus also spoke of the transmission of errors from older manuscripts to newer ones, and, what is more, he seems to be the first to speak of this phenomenon in genealogical terms:

> The agreement of manuscripts will not seem at all astonishing to those who have even a modicum of experience in assessing and collating manuscripts. For it very often happens that an error of one archetype, so long as it has some specious appearance of the truth, goes on to propagate itself in all the books that form, as it were, its descendants, "and the children of its children and those who are born later."[21]

Thus Erasmus was quite aware that the sins of an archetype could be visited on its copies and their copies. Furthermore, Erasmus here discusses errors that may have "some specious appearance of the truth," so that they may masquerade as legitimate characteristics and, therefore, be "propagated" without detection in more and more descendants, like a hereditary mark or contagion.

The French philologist Joseph Scaliger (1540–1609) extended this metaphor of bastardized descendants to include the compilations of variants (*adversaria*) produced by the textual critics of his day:

> We have made many observations on authors of both languages [Greek and Latin] from which could arise a monstrous progeny of Variae, Antiquae Lectiones, Miscellanea, and other such things in which the ambition of Philologists today wantonly indulges.[22]

In this metaphor, the philologists, by publishing collections of readings, breed ("wantonly indulge") with diverse texts to give birth to "monstrous progeny." Scaliger, on the other hand, wished to do away with erroneous variants altogether:

---

21 Erasmus, *Adagiorum Chiliades* I.6.36 (Basel: Froben, 1538); cited and translated by Timpanaro, *Genesis*, 49. The quotation at the end is from Homer, *Iliad* 20.308 (see Timpanaro, ibid., n. 16).

22 Joseph Scaliger, *Epistolae* (Leiden: Bonaventura and Abraham Elzevir, 1627), 52, in a letter to Janus Dousa: "Multa in auctores utriusque linguae observavimus, ex quibus immanes partus nasci possent Variarum, Antiquarum Lectionum, Miscellaneorum, & aliorum eiusmodi, quibus hodie Philologorum ambitio lasciuire solet." Cited by Timpanaro, *Genesis*, 51 n. 19.

> So that we might obtain the fruit of our efforts, we undertook to interpret and purify whole authors.[23]

Scaliger hoped to detect and remove the flaws of the descendants in order to obtain the archetype.[24] One observation Scaliger made of ancient manuscripts was that the type of script used by the scribe was a significant element to consider in textual reproduction: the "crabby characters" of "Langobardian letters" of a manuscript of Catullus, for example, might easily be misread and miscopied, thus giving birth to corrupt offspring.[25] Through this observation, Scaliger attempted to reconstruct the history of transmission and, thus, the archetype, by tracing a genealogy of error.[26]

Nevertheless, the use of genealogical tables to illustrate the relationships between texts was not proposed until Bengel. It is unsurprising that a New Testament textual critic rather than a Classical one should conceive of such a plan, because of the relative wealth of biblical manuscripts in Bengel's time (and now numbering in the thousands) in contrast to Classical texts—for example, Lucretius's *De rerum natura,* of which there are only four extant manuscripts older than the ninth century.[27] As discussed in the previous chapter, Bengel's methodology for managing such great numbers of readings was classification into taxonomic groups, and then the evaluation of these groups based on their genealogical ties. But Bengel did not fulfill his plans for the *schematismos* he envisioned, and it is to Lachmann that credit must be given as the one who guided New Testament textual criticism toward the genealogical method of *recensio.*

---

23 Ibid.: "Ut tamen nobis vigiliarum nostrarum fructus constaret, interpretandos et emaculandos auctores integros suscepimus." Cited by Timpanaro, *Genesis,* 51 n. 19.

24 See Reynolds and Wilson, *Scribes and Scholars,* 176 and Timpanaro, *Genesis,* 51–2.

25 See Timpanaro, *Genesis,* 51 n. 20.

26 See Reynolds and Wilson, *Scribes and Scholars,* 176: "In his edition of Catullus he sought to prove from the nature of the corruptions in the manuscripts (such as the confusion of open *a* and *u*, long *i* and *l*) that they were descended from a common parent in what he called 'Lombardic' script; by this he appears to have meant something like the Visigothic hand he had encountered when working on Ausonius. Though he was wrong, he had gone further than anyone in trying to reconstruct the details of a lost archetype and in making the history of a particular text an important criterion in establishing it."

27 See Michael Reeve, "Lucretius in the Middle Ages and Early Renaissance: Transmission and Scholarship," in *The Cambridge Companion to Lucretius,* ed. Stuart Gillespie and Philip Hardie (Cambridge: Cambridge University Press, 2007), 205–6.

In the preface of his 1842 edition of the Greek New Testament, Lachmann articulates the different methods and aims of *recensio* and *emendatio*, placing *recensio* first and foremost as the task of the textual critic:

> To present the writings of the ancients we make use of two skills: we both discuss which scribe wrote what and lay out what he meant and thought in given circumstances; the one skill is claimed by critical ability; the other consists of interpretation. There are three steps of criticism: recension, emendation, and detection of the origin. What was written is perceived in two ways: by examining the witnesses and by restoring their testimony, where it errs, to the truth: in this way one progresses gradually from the writings to the writer. Thus, we must investigate above all what the most reliable scribes handed down; then we must determine what could have come from the writer's hand; in the third step, we must explore who wrote at what time, under what circumstances, and using which aids. Investigating the scribes, which I placed first, what is called recension, is something that we both can and should do without use of interpretation; interpretation, on the other hand, cannot be applied unless what the witnesses report is understood, cannot be accomplished unless the writer has been identified. And so emendation and the investigation of the origin of the book, because it pertains to recognizing the mind of the writer, uses interpretation as its foundation. Thus no part of this task can be separated from the others safely, except that one which should be foremost: I mean that which examines the reliability of the witnesses and which presents what the most trustworthy scribes have transmitted.[28]

28 Karl Lachmann, "Praefatio," in *Novum Testamentum Graece et Latine*, vol. 1 (Berlin: Reimer, 1842), v: "Ad scripta veterum repraesentanda duabus diversis utimur artibus: nam et qui scriptor quid scripserit disputamus, et quo rerum statu quid senserit et cogitarit exponimus; quorum alterum sibi iudicandi facultas vindicat, alterum interpretatione continetur. iudicandi tres gradus sunt, recensere, emendare, originem detegere. nam quid scriptum fuerit, duobus modis intellegitur, testibus examinandis, et testimoniis, ubi peccant, revocandis ad verum: ita sensim a scriptis ad scriptorem transiri debet. itaque ante omnia quid fidissimi auctores tradiderunt quaerendum est, tum quid a scriptoris manu venire potuerit iudicandum, tertio gradu quis quo tempore qua condicione quibus adminiculis usus scripserit explorandum. ex auctoribus quaerere, quod primo loco posui, id quod recensere dicitur, sine interpretatione et possumus et debemus: contra interpretatio, nisi quid testes ferant intellectum fuerit, locum habere, nisi de scriptore constiterit, absolvi non potest: rusrus emendatio et libri originis investigatio, quia ad ingenium scriptoris cognoscendum pertinet, tamquam fundamento nititur interpretatione. quo fit ut nulla huius negotii parts tuto a ceteris separari possit, nisi illa una quae debet esse omnium prima: illam dico quae testium fidem perscrutatur et locupletissimis auctoribus tradita repraesentat."

Thus Lachmann prizes most of all *recensio*, what he believes to be the foundation of all other work by the textual critic, and a task he believes possible *sine interpretatione*. *Recensio*, in Lachmann's understanding, is comprised of determining "which scribe wrote what," "examining the witnesses," and "[investigating] above all what the most reliable scribes handed down." In other words, *recensio* is examining and separating out different strands of textual transmission, and determining which strands are the most trustworthy. Only after this has been done could the work of *emendatio* begin, since working to correct the errors of an unreliable text would be useless. This also meant the acceptance of the content of what were deemed reliable witnesses regardless of how they might offend a philologist's sense of style or grammar.

This further meant the rejection of the *Textus Receptus*, a work built upon *emendatio*. In Lachmann's opinion, even Griesbach, who departed from the received text at times, and Bengel, whom he praised as truly understanding the work of a textual critic, did not go far enough. Of course Lachmann had nothing but disdain for those who insisted upon reproducing the *Textus Receptus* and using it as the basis of their investigation of the witnesses:

> What shall we say about the intentions of those who scorn precisely this first step and refuse to criticize the books of the New Testament? That it seems more correct to them to review the common, vulgate edition and where possible emend it. Really? They prefer the work of chance, of carelessness, of negligence to diligent and sustained labor?[29]

By dismissing the received text, Lachmann was free to determine *sine interpretatione* (so he believed) which texts were the most reliable, and this was done with consideration of their relationships to one another, whether near or distant both in readings and in geographic location:

> There is nothing that can be better attested than that in which all authorities from everywhere agree. Less effective is the consensus

29 Ibid.: "Quid igitur eos spectare dicemus, qui hunc ipsum primum gradum contemnentes libros novi testamenti recensendos negant? scilicet illis rectius agere videtur qui communem et pervulgatam editionem recognascat et ubi fieri possit emendet. itane vero? navo et continuo labori operam fortuitam remissam neglegentem praeferunt?"

> if part of them is silent or in some way defective. Moreover the strength in the harmony of witnesses gathered from different areas is greater than the danger posed by some kindred witnesses that disagree with them either because of negligence or as if by design. But also ambiguous testimony should be evaluated, when the consensus of remote witnesses is weakened by the consensus of other witnesses, also from disparate places, with different testimony. Uncertain testimony is that which is different in different regions but which has been transmitted consistently in individual regions.[30]

Lachmann did not trace out a genealogy of manuscripts for the New Testament, but he laid the groundwork so that it became the next logical step: if determining the most trustworthy texts was *the* most important task of criticism, and this was done by comparing their different readings with those from other groups, then mapping out their relationships, their *kinship*, was the next desideratum.

It is his work in classical philology, his application of the genealogical method to the texts of Lucretius, that brought the use of *stemmata codicum* to the forefront of textual criticism.[31] Lachmann's work on Lucretius not only traced the relationships of surviving texts back to the archetype, the exemplar from which all copies descended, but also claimed to reconstruct even the format, down to the precise number of lines per page, of that original text. This introduced a new holy grail to the world of textual criticism, especially for New Testament textual criticism: the attainment of the archetype could be equated with discovering the very words of God.

This aim put the elements of the genealogical method into sharp relief: those surviving witnesses available to textual critics were a means to an

30 Ibid., viii: "itaque nihil testatius esse potest eo in quo omnes undique auctores consentiunt. paulo minus efficit consensio, si eorum pars taceat vel aliquo modo deficiat. porro maius in testium ex diversis regionibus coactorum concordia firmamentum est quam ab aliquot popularibus vel neglegentia vel quasi compacto ab illis discedentibus periculum. sed ancipitia censenda sunt testimonia, cum longinquorum testium consensus aliorum item locis diiunctorum in diversa testificatione consensu elevatur. incerta sunt quae in aliis regionibus alia, sed constanti singularum consuetudine, tradita sunt. postremo infirma, in quibus ne eaedem quidem regiones consentiunt."

31 On Jacob Bernay's work on Lucretius *prior* to Lachmann's edition, which also made use of genealogical stemmata, see Timpanaro, *Genesis*, 102–6. On Lachmann as the innovator of the genealogical method, see Timpanaro, *Genesis*, 115: "The true founders of the genealogical classification of manuscripts are Schlyter (in a field different from Classical philology), Zumpt, Madvig, and above all Ritschl and Bernays, who concretely fulfilled the requirement to which Bengel had given expression."

end, their singular readings most likely errors, and these errors were to be used in mapping out a family tree of texts that would lead to an ancestor, and the ancestor was more important than anything else. The witnesses, the "monstrous progeny," were corrupt, the source pure, and it was backwards to the source that all criticism should lead.

This preoccupation with origins was, of course, not unique to the field of textual criticism. During the eighteenth and nineteenth centuries, in the interest of scientific inquiry into all facets of life, all layers of folklore, story, myth, and belief were to be peeled back to reveal the truth. Whether it was determining the root of all religious ritual or the investigation of the historical Jesus, the gaze of the scholar looked ever toward the past.[32] Thus Westcott and Hort describe textual criticism in this way:

> Its progress consists not in the growing perfection of an ideal in the future, but in approximation towards complete ascertainment of definite facts of the past, that is, towards recovering an exact copy of what was actually written on parchment or papyrus by the author of the book or his amanuensis.[33]

Genealogies and the tracing of genealogies do not necessarily presuppose a concern for purity, mixture, and contamination. But in the field of textual criticism, where "descendants" in a family line are actual copies of an ancestor, it seems logical to assume that the "purer" a copy, the better. Sebastiano Timpanaro provides a useful comparison between textual criticism and linguistics, both fields that use genealogies to map their subject matter:

> There is an undeniable affinity between the method with which the Classical philologist classifies manuscripts genealogically and

---

32 On eighteenth- and nineteenth-century investigations of the origins of ancient religions and their comparison to an "authentic" Christianity that was supposedly *sui generis*, see Jonathan Z. Smith, *Drudgery Divine: On the Comparison of Early Christianities and the Religions of Late Antiquity* (Chicago: University of Chicago Press, 1990). On the search for the "primordial" and "authentic" in biblical scholarship of the twentieth and twenty-first centuries, see Shawn Kelley, "Race, Aesthetics, and Gospel Scholarship: Embracing and Subverting the Aesthetic Ideology," in *Prejudice and Christian Beginnings: Investigating Race, Gender, and Ethnicity in Early Christian Studies*, ed. Laura Nasrallah and Elisabeth Schüssler Fiorenza (Minneapolis, MN: Fortress Press, 2009), 191–209.

33 "Introduction," in *The New Testament in the Original Greek*, ed. Brooke Foss Westcott and Fenton John Anthony Hort (Cambridge: Macmillan, 1882), 3.

> reconstructs the reading of the archetype, and the method with which the linguist classifies languages and as far as possible reconstructs a lost mother language, for example, Indo-European. In both cases inherited elements must be distinguished from innovations, and the unitary anterior phase from which these have branched out must be hypothesized on the basis of the various innovations. . . . Naturally, like all analogies, this one too is valid only within certain limits: in linguistics there is nothing corresponding to the distinction between *archetype* and *original*; and even when a corruption has spread through the whole manuscript tradition, it is still always felt to be something that disturbs the context (and hence it is felt as a real error, or as a banalization that constitutes a deterioration too, not as a simple neutral "innovation"), while a linguistic innovation, once it has achieved success, ceases by that very fact to be felt as an error.[34]

For the textual critics of the eighteenth and nineteenth centuries, the value of witnesses lay in their preservation of the original archetype or autograph. Where a reading was seen as a deviation from the autograph, it was not simply part of an evolutionary process of change and growth but a failure. Its only value lay in the evidence it provided for the *stemma codicum*. Otherwise, it was a corruption and, what is worse, a corruption that had almost undoubtedly infected its descendants.

The racialization of texts through classification paved the way for the equation of variants with corruption and contamination. By dividing the witnesses into "families, tribes, and nations," textual critics introduced the possibility of "mixture" between these groups. Once they abandoned the *Textus Receptus* and began the quest for the original archetype, the characteristics of the "nations" and the offspring born to them became marks of difference, and differences from the progenitor became marks of corruption.

This chain of consequential understandings of manuscripts and their differences paralleled contemporary biological understandings of the human species once it was racialized and classified. The classification of the human species by race began in earnest with Linnaeus, as discussed

34 Timpanaro, *Genesis*, 119. For a comparison of the genealogical methods in linguistics and the developments in the natural sciences, see Henry M. Hoenigswald and Linda F. Wiener, eds., *Biological Metaphor and Cladistic Classification: An Interdisciplinary Perspective* (Philadelphia: University of Pennsylvania Press, 1987).

in the previous chapter, but the concepts of human classification and the "mixture" of different human "types" predates Linnaeus and his successors. Just as the Humanists Politian, Erasmus, and Scaliger were precursors to Bengel and later Lachmann in terms of understanding texts and manuscripts as related genealogically, so also there were precursors to racial classification and the understanding of racial and genealogical "purity" in the late Middle Ages and the Renaissance.

The concept of "cleanliness of blood," *limpieza de sangre*, as something to be proven by genealogical records was first enacted in Toledo, Spain, in 1449.[35] This was at the tail end of the Reconquista, when all Muslims and Sephardic Jews were soon to be expelled from the country (in 1492). Thus, new converts to Christianity were viewed with suspicion by "Old Christians," who blamed economic woes on *conversos* who had taken political and church positions.[36] The statutes of *limpieza de sangre* required that candidates for official positions prove they were not of Moorish or Jewish extraction. Here, of course, the focus was on religious identification, but that identification was only legitimate if it could be proven to have been the case for several generations, thus linking religion (a system of classification) to genealogy and "purity of blood."

A likely first use of the term "race" to designate partitions (or breeds) within a species occurs in the fifteenth century in the poem *The Hunt* by the seneschal Jacques de Brézé to address a "race" of dogs,[37] and it is with the breeding of hunting animals (such as dogs and falcons) that the notion of "noble race" comes to the fore in the tracing of genealogies.[38] As for subdividing the human species, the inspiration for Linnaeus's

---

35 See Gregory B. Kaplan, "The Inception of *Limpieza de Sangre* (Purity of Blood) and Its Impact in Medieval and Golden Age Spain," in *Marginal Voices: Studies in* Converso *Literature of Medieval and Golden Age Spain*, ed. Amy Aronson-Friedman and Gregory B. Kaplan (Leiden: Brill, 2012), 19–41. See also the classic work on the statutes, Albert A. Sicroff, *Les controverses des status de "pureté de sang" en Espagne du XVe au XVIIe siècle* (Paris: Didier, 1960).

36 Kaplan, "Inception of the *Limpieza de Sangre*," 26.

37 Jacques de Brézé, Le livre de la chasse du Grand Seneschal de Normandye et Les ditz du bon chien Souillard (Paris: Aubry, 1858).

38 Charles de Miramon, "Noble Dogs, Noble Blood: The Invention of the Concept of Race in the Late Middle Ages," in *The Origins of Racism in the West*, ed. Miriam Eliav-Feldon, Benjamin Isaac, and Joseph Ziegler (New York: Cambridge University Press, 2009), 200–16.

categories of *Homo sapiens* was François Bernier (1620–1688), who divided the human species into four types, one from each of the four quarters of the earth.[39] Linnaeus's racial classification of humans, however, systematized the notions of difference (skin color) and geography (by Europe, Asia, Africa, and America) in the context of a *totalizing* taxonomy of all living organisms. This set the stage for all subsequent racial classification, just as Bengel's Asiatic and African text "races" did for New Testament textual criticism.

Classification of both texts and humans raised several questions: if texts and humans were divided into races, could individuals change from one race to another? What would happen if individuals from different races produced offspring? What if these mixed offspring continued to mate with others like themselves?

As mentioned in the last chapter, Buffon promoted the theory of species degeneration, claiming that each species contains a *moule intérieure* from which it can degenerate but to which it can also return, given the right circumstances. If a degenerate group returned to its native climate and environmental context, it could gradually, over hundreds of years, be restored to its original form. However, a faster way to return a group to its *moule intérieure* would be through mixture:

> [If] a man should be obliged to forsake those climates which he has possessed himself of, and return to his native country, he would in time resume his original features, his primitive size, and his natural colour. But the mixture of races would produce the same effect in a much shorter time. The conjunction of a White male with a Black female, or a Black male with a White female, equally produce a mulatto, whose colour is brown, that is, a mixture of Black and White. The mulatto intermixing with a White produces a second mulatto not so brown as the former; and if this second mulatto unites with a White, the third mulatto will have no more than a slight tincture of the brown, which will entirely disappear in succeeding generations. Thus, by the mixture with a White, one hundred and fifty, or two hundred years is sufficient to bleach the skin of the Negro; but it would,

39 C. Loring Brace, *"Race" Is a Four-Letter Word: The Genesis of the Concept* (New York: Oxford University Press, 2005), 27. See also Siep Stuurman, "Francois Bernier and the Invention of Racial Classification," *History Workshop Journal* 50 (2000): 1–21.

> perhaps, require many centuries to produce this effect by the influence of climate alone.[40]

Buffon does not state explicitly that the *moule intérieure* from which human races can degenerate is represented by the White "race," but the example given above to demonstrate a return to a "native color" hypothesizes the "bleaching" of Negro skin through repeated mating until the resultant offspring is White. In any case, the original interior mold can be corrupted and adulterated, just like an archetype in textual criticism, and implied in Buffon's example is that the White race is that ancestral race.

Writing shortly after Buffon, Blumenbach specifically identifies the "Caucasian race" as the progenitor of the four other races of his classification, the Mongolian, Ethiopian, American, and Malay:

> I have allotted the first place to the Caucasian, for reasons given below, which make me esteem it the primeval one. This diverges both directions into two, most remote and very different from each other; on the one side, namely, into the Ethiopian, and on the other into the Mongolian. The remaining two occupy the intermediate positions between that primeval one and these two extreme varieties; that is, the American between the Caucasian and Mongolian; the Malay between the same Caucasian and Ethiopian.[41]

---

40 Translation by J. S. Barr, *Buffon's Natural History* (London: Gillet, 1807), 9: 318, with modifications. Buffon, "De la dégénération des animaux," in *Histoire naturelle, générale et particulière, avec la description du Cabinet du Roy*, vol. 14 (Paris: L'Imprimerie Royale, 1766), 313: "s'il arrivoit, dis-je, que l'homme fût contraint d'abandonner les climats qu'il a autrefois envahis pour se réduire à son pays natal, il reprendroit avec le temps ses traits originaux, sa taille primitive et sa couleur naturelle : le rappel de l'homme à son climat amèneroit cet effet, le mélange des races l'amèneroit aussi et bien plus promptement; le Blanc avec la Noire, ou le Noir avec la Blanche produisent également un Mulâtre dont la couleur est brune, c'est-à-dire, mêlée de blanc et de noir; ce Mulâtre avec un Blanc produit un second Mulâtre moins brun que le premier; et si ce second Mulâtre s'unit de même à un individu de race blanche, le troisième Mulâtre n'aura plus qu'une nuance légère de brun qui disparoîtra tout-à-fait dans les générations suivantes : il ne faut donc que cent cinquante ou deux cents ans pour laver la peau d'un Nègre par cette voie du mélange avec le sang du Blanc, mais il faudroit peut-être un assez grand nombre de siècles pour produire ce même effet par la seule influence du climat."

41 Translated by Thomas Bendyshe in Johann Friedrich Blumenbach, *On the Natural Varieties of Mankind* (New York: Bergman Publisher, 1969), excerpted in Eze, *Race and the Enlightenment*, 84. Johann Friedrich Blumenbach, "Generis humani varietates quinae principes, species vero unica," in *De generis humani varietate nativa*, 3rd ed. (Göttingen: Vanderhök and Ruprecht, 1795), Sectio IV §81, p. 286–7: "*Caucasiam* ob caussas infra enarrandas pro primigenia habendam primo loco posui. Haec utrinque in bina ab invicem

According to Blumenbach, since the Caucasian race presented the most perfect and beautiful form of the species, it was from this stock that all the other races degenerated.[42]

Blumenbach did not intend to imply that degenerated groups were necessarily intellectually or morally inferior, but the concept of degeneration, influenced by both his and Buffon's scientific work, became a major preoccupation of European scientists and social pundits. From the nineteenth to the twentieth century, Europe was gripped with the fear that certain members of its population had degenerate moral and mental capacities and were naturally criminal and sexually perverse.[43]

At the same time, European and American imperialistic enterprises brought the question of racial mixture to the fore: if colonists or slaveholders produced offspring with non-whites, what would be the consequences? After having separated and classified the races, scientists were now hypothesizing the results of their mixture. Since they had posited a pure, original, primary stock, it was only a matter of course that "mixed" offspring would be impure and degenerative. In the case of race, impurities were not only outwardly expressed through behavior but were also transmitted by blood. As Buffon states, degenerative humans were not a different species but degenerations were carried in the blood: "The blood is different, but the germ is the same."[44]

Thus a concern for origins and purity went hand in hand with paranoia over the corruption of a race, thought to be pure, via the transmission of flaws through the blood. So Gobineau writes:

> The word *degenerate*, when applied to a people, means (as it ought to mean) that that people no longer has the same intrinsic value as it had before, because it has no longer the same blood in its veins,

remotissima et diversissima extrema abiit, hinc nempe in *Mongolicam*, illinc in *Aethiopicam*. Medios vero inter istam primigeniam et hasce extremas varietates locos tenent reliquae binae: *Americana* nempe inter Caucasiam et Mongolicam. *Malaica* inter eandem istam Caucasiam et Aethiopicam."

42 See Blumenbach, Sectio IV § 85, p. 303: "A) Varietas Caucasia. Nomen huic varietati a Caucaso monte, tum quod vicinia eius et maxime quidem australis plaga pulcherrimam hominum stirpen, Georgianam foveat; tum quod et omnes physiologicae rationes in eo conspirent in eandem regionem, si uspiam, primos humani generis autochthones verisimillime ponendos esse."

43 See Daniel Pick, *Faces of Degeneration: A European Disorder c. 1848-c. 1924*, Ideas in Context 15 (Cambridge: Cambridge University Press, 1989).

44 Buffon, "De la dégénération des animaux," 313: "le sang est différent, mais le germe est le même."

> continual adulterations have gradually affected the quality of the blood. In other words, though the nation bears the name given by its founders, the name no longer connotes the same race; in fact, the man of a decadent time, the *degenerate* man properly so called, is a different being, from the racial point of view, from the heroes of the great ages. I agree that he still keeps something of their essence; but the more he degenerates the more attenuated does this "something" become. The heterogeneous elements that henceforth prevail in him give him quite a different nationality—a very novel one, no doubt, but such originality is not to be envied. He is only a very distant kinsman of those he still calls his ancestors.[45]

The mixture of different races here is presented as a dilution of something to its detriment, so that a once great race devolves into weakness and decadence. Gobineau also hints at an insidious threat posed by interracial progeny, who can appear to be of the same race as their illustrious ancestors and may go by the same name, but who are imposters because they are something else altogether.

Here, as in textual criticism, novelty is not innovation but disruption of purity and lineage. Impurities introduced into the bloodline pollute it and multiply through the generations, just as variant readings are introduced into a text and then reproduced. So Westcott and Hort write:

> Every transcription of any kind of writing involves the chance of the introduction of some errors. . . . When the transcript becomes itself the parent of other copies, one or more, its errors are for the most part reproduced. . . . In addition to inherited deviations from

45 Translation by Adrian Collins from Arthur de Gobineau, *The Inequality of the Human Races* (New York: Putnam, 1915), 25, with modifications. Joseph Arthur Comte de Gobineau, *Essai sur l'inégalité des races humaines* (Paris: Didot, 1853), 39–40: "Je pense donc que le mot dégénéré, s'appliquant à un peuple, doit signifer et signifie que ce peuple n'a plus la valeur intrinsèque qu'autrefois il possédait, parce qu'il n'a plus dans ses veines la même sang, dont des alliages successifs ont graduellement modifié la valeur; autrement dit, qu'avec le même nom, il n'a pas conservé la même race que ses fondateurs; enfin, que l'homme de la décadence, celui qu'on appelle l'homme dégénéré, est un produit différent, au point de vue ethnique, du héros des grandes époques. Je veux bien qu'il possède quelque chose de son essence; mais, plus il dégénère, plus ce quelque chose s'atténue. Les éléments hétérogènes qui prédominent désormais en lui composent une nationalité toute nouvelle et bien malencontreuse dans son originalité; il n'appartient à ceux qu'il dit encore être ses pères, qu'en ligne très-collatérale."

> the original, each fresh transcript is liable to contain fresh errors, to be transmitted in like manner to its descendants.[46]

The genealogical chart of texts grows and multiplies in impurities when texts are copied from multiple exemplars, different manuscripts that mingle and mate, creating a "mixture," so that their errors—and accuracies—are blended in a new hybrid:

> The gain or loss to the intrinsic purity of texts from mixture with other texts is from the nature of the case indeterminable. In most instances there would be both gain and loss; but both would be fortuitous, and they might bear to each other every conceivable proportion.[47]

This is similar to Buffon's description of interbreeding amongst races: a relatively corrupt text can, with mixture, become purer, just as a relatively pure text can become further corrupted. Westcott and Hort's evaluation of *loss* of purity as "fortuitous" must be placed in the context of genealogical textual criticism, in which impurities serve as the basis for genealogical reconstruction and, therefore, as crucial evidence for uncovering the untainted original, the primary goal of modern text critics.

That this was *not* the primary goal of ancient transcribers and users of the New Testament is addressed immediately after the above quotation:

> Textual purity, as far as can be judged from the extant literature, attracted hardly any interest. There is no evidence to shew that care was generally taken to choose out for transcription the exemplars having the highest claims to be regarded as authentic, if indeed the requisite knowledge and skill were forthcoming. Humanly speaking, the only influence which can have interfered to an appreciable extent with mere chance and convenience in the selection between existing readings, or in the combination of them, was supplied by the preferences of untrained popular taste, always an unsafe guide in the discrimination of relative originality of text.[48]

46 Westcott and Hort, "Introduction," 5.

47 Ibid, 8–9.

48 Ibid., 9.

Yet despite their belief that those closest in time to the sought-after original text had little concern for its purity, Westcott and Hort state without irony and as a matter of fact that "textual criticism is always negative, because its final aim is virtually nothing more than the detection and rejection of error."[49] In the stark world of purity and corruption, the histories of "popular taste" and of varying scribal and communal values were of no worth except in terms of their usefulness for discovering the unstained original.

Clearly what began as a metaphor for texts and their characteristics came to shape the very goals of textual criticism, indeed, to determine the very definition of "truth," especially for New Testament scholars. Regarding his critical edition of the Greek New Testament, Tregelles writes:

> and from [the oldest copies], in connection with other ancient evidence, it is for criticism, in a Christian spirit and with proper intelligence, to seek the materials for reconstructing that fabric of revealed truth which has been in some measure disfigured by the modern copyists and their followers, the early printers.[50]

And in a similar vein Henry Alford praises Lachmann for his "bold and uncompromising demolition of the unworthy and pedantic reverence for the received text, which stood in the way of all chance of discovering the genuine word of God."[51]

Lachmann, in his embrace of Bengel's classification and proposal for a genealogy of texts, succeeded in toppling the *Textus Receptus*. He wrote with scorn regarding those who would prefer "elegance" over "truth,"[52] and, in keeping with biological trends, equated truth with a pure and uncontaminated progenitor, the ancestor of all textual descendants. In short, he replaced the *Textus Receptus* with the "original text," and the defense of a received scripture with the detection and elimination of degeneration.

---

49 Ibid., 3.

50 Samuel Prideaux Tregelles, *An Account of the Printed Text of the Greek New Testament* (London: Bagster, 1854), 181–2.

51 Henry Alford, *The Greek New Testament*, 5th ed., vol. 1 (London: Rivingtons, 1863), 76; partially quoted in Metzger and Ehrman, *Text of the New Testament*, 174.

52 See Karl Lachmann, ed., *Sextus Aurelius Propertius: Carmina* (Leipzig: Fleischer, 1816), xvii–xviii; cited in Timpanaro, *Genesis*, 77.

Conclusion

From the organization of diversity by classification to the hierarchization of groups and finally to the association of variation with impurity and contamination, textual criticism kept in step with the developments of natural science, adopting its language and borrowing its structures. The race ideology of the eighteenth and nineteenth centuries led to legislation in Europe and America, and their colonial holdings, against miscegenation, immigration, and participation of certain "races" in political bodies, all in the interest of protecting the idealized pure race with which these nation-states identified.[53] The concept of *Herrenvolk*, a "master race," in Germany would reach its devastating climax in the twentieth century. After the Nuremberg Trials, the parlance of pure blood and blood taint would drop out of biological discourses of heredity,[54] but the vestiges of eighteenth- and nineteenth-century race ideology remain detectable in the metaphors of textual criticism through terms such as "textual complexion," and "textual physiognomy," as well as genealogical terms of degeneration such as "corruption," "cross-contamination," and "mixed ancestry."[55]

The great irony of the classification and subsequent hierarchization of both human races and text families in the eighteenth and nineteenth centuries is that no comprehensive genealogy of difference was ever produced to provide evidence for the histories of corruption and degeneration

---

53 On concerns over miscegenation and "blood taint" in South Africa and Europe at the beginning of the twentieth century, see John Maxwell Coetzee, "Blood, Flaw, Taint, Degeneration: The Case of Sarah Gertrude Millin," *English Studies in Africa* 23 (1980): 41–58. On American legislation against the legal representation of "Indians" and "descendants of Africans," see Priscilla Wald, "Terms of Assimilation: Legislating Subjectivity in the Emerging Nation," in *Cultures of United States Imperialism*, ed. Amy Kaplan and Donald E. Pease (Durham, NC: Duke University Press, 1993), 59–84. On American anti-immigration and anti-miscegenation laws aimed at Asians and Asian Americans, see Susan Koshy, *Sexual Naturalization: Asian Americans and Miscegenation* (Stanford: Stanford University Press, 2004) and Lisa Lowe, *Immigrant Acts: On Asian American Cultural Politics* (Durham, NC: Duke University Press, 1996).

54 Coetzee, "Blood, Flaw, Tain, Degeneration," 41.

55 The use of "corruption" and "contamination" in current New Testament textual criticism is ubiquitous. A couple of notable examples are Gerd Mink, "Problems of a Highly Contaminated Tradition: The New Testament; Stemmata of Variants as a Source of Genealogy for Witnesses," in *Studies in Stemmatology II*, ed. Pieter van Reenan, August den Hollander, and Margot van Mulken (Amsterdam: John Benjamins, 2004), 13–85 and (rather more tongue-in-cheek) Bart D. Ehrman, *The Orthodox Corruption of Scripture: The Effect of Early Christological Controversies on the Text of the New Testament* (Oxford: Oxford University Press, 2003). Metzger and Ehrman use both "cross-contamination" and "mixed ancestry" in *Text of the New Testament*, 106, 208.

hypothesized. Westcott and Hort, despite their compelling arguments for the use of the genealogical method to dethrone the *Textus Receptus* and posit a stemma of text families, never used the method to prove their conclusions. Ernest Colwell, writing roughly sixty-five years after the publication of Westcott and Hort's edition, states the facts with characteristic acumen:

> That Westcott and Hort did not apply this method to the manuscripts of the New Testament is obvious. Where are the charts which start with the majority of late manuscripts and climb back through diminishing generations of ancestors to the Neutral and Western Texts? The answer is that they are nowhere.[56]

In fact, not only did Westcott and Hort not use the method to provide a stemma of texts, neither did any other scholar throughout most of the twentieth century:

> Yet genealogical method as defined by Westcott and Hort was not applied by them or by any of their followers to the manuscripts of the New Testament. Moreover, sixty years of study since Westcott and Hort indicate that it is doubtful if it can be applied to New Testament manuscripts in such a way as to advance our knowledge of the original text of the New Testament.[57]

The reasons for the absence of comprehensive genealogies of both human races and New Testament texts are the same: first of all, the overwhelming amount of data could not, before the age of computers, be managed and analyzed to produce global stemmata; second, no underlying theory of inheritance existed in either biology or philology to help adjudicate between two direct offspring (siblings) to determine the nature of the missing parent;[58] third, and likewise, no underlying theory of inheritance existed to account for "mixture" between parents—and Colwell states as much in

---

56 Ernest C. Colwell, "The Genealogical Method: Its Achievements and Its Limitations," *Journal of Biblical Literature* 66 (1947): 112.

57 Ibid., 109–10.

58 See ibid., 113: "The nature of this limitation can be easily illustrated. Common ancestry is indicated by agreement in reading; if there are two brothers and a cousin, the agreement of one brother with the cousin will establish the parental nature even where the brothers differ; but if the two brothers have no cousins and disagree, genealogy can assert nothing as to which trait comes from the father."

exactly the same metaphorical terms: "The intermarriage of manuscripts knows no more restrictions than exist in the human race."[59]

Colwell, immediately above this last quotation, gives an analogy of human families just as Gordon Fee did more than twenty years later, to illustrate the challenges of the method:

> The difficulty that confronts the user of genealogical method in mixed texts may be paraphrased in human terms thus. My mother was of German origin through the "Pennsylvania Dutch." My father was born in the South of England. My wife's mother is of French descent through the Huguenots who settled in the Carolinas. Her father was born in the North of England. We have two children. By studying them, could you reconstruct us; or if we were missing—as is usually the case in New Testament manuscripts—could you reconstruct our parents and so on back to Adam? When the children agree in some British trait, are they agreeing with us or with their grandparents or with some proto-Briton?[60]

For the field of biology, however, the seeds of change were already planted in the middle of the nineteenth century, by Darwin and Mendel. Together, their theories of natural selection and heredity would become the foundation of evolutionary biology and genetics in the twentieth century and beyond, providing the theories of inheritance necessary to map out, comprehensively, both the evolutionary relationships of all living organisms on the planet as well as the entire genetic makeup of a human being.

Textual criticism would follow suit slowly, first moving away from the understanding of variants as simply "corruption" toward seeing them as also a part of the evolutionary history of the textual tradition. Difference, rather than viewed as degeneration, would, in a Darwinian perspective, be understood as interesting in and of itself as a stage in a living text's adaptation to its environment. Additionally, at the turn of the twenty-first century, a global stemma of the evolutionary relationships of all extant texts would be within reach through advances in theory and computer software. New Testament textual critics could move beyond classifying

59 Ibid., 115.

60 Ibid.

texts and, instead, build comprehensive, global stemmata so that the relationships between child and parent and breeding pairs of manuscripts could be illustrated in great detail. By the latter half of the twentieth century, New Testament textual criticism would, for the most part, leave the methods of classification and rudimentary genealogy behind.

PART II

Historicization and Innovation

3

Darwin, Streeter, and Narrative Textual Criticism

But the book which has most engaged me is Darwin. Whatever may be thought of it, it is a book that one is proud to be contemporary with. I must work out and examine the argument more in detail, but at present my feeling is strong that the theory is unanswerable. If so, it opens up a new period—I know not what.

F. J. A. Hort, in a letter to the Rev. John Ellerton, April 3, 1860

For a polymath like Hort, the year 1860 could not pass without serious discussion and debate about two highly controversial publications in the fields of science and theology, respectively Darwin's *On the Origin of Species* and a collection of articles by seven English churchmen titled *Essays and Reviews*.[1] The latter, published in March of 1860, and following hot on the heels of the *Origin of Species*, published in November of 1859, represented the Anglican Broad Church movement and challenged traditional views in its call for scientific and historical critical readings of the bible.[2] In one of the essays, "On the Study of the Evidences of Christianity," Baden Powell would give the first positive assessment by a theologian of Darwin's monumental work:[3]

1 The epigraph comes from Fenton John Anthony Hort, *Life and Letters of Fenton John Anthony Hort*, ed. Arthur Fenton Hort (London: Macmillan, 1896), 416. Charles Darwin, *On The Origin of Species by Means of Natural Selection* (London: Murray, 1859); *Essays and Reviews* (London: John W. Parker and Son, 1860).

2 For a history of its reception and the furor of counterattacks and ecclesial court trials (as well as a carefully edited version of the articles themselves), see *Essays and Reviews: The 1860 Text and Its Reading*, ed. Victoria Shea and William Whitla (Charlottesville, VA: University Press of Virginia, 2000).

3 Shea and Whitla, *Essays and Reviews*, 272 n. 88.

> a work has now appeared by a naturalist of the most acknowledged authority, Mr. Darwin's masterly volume on *The Origin of Species* by the law of "natural selection,"—which now substantiates on undeniable grounds the very principle so long denounced by the first naturalists,—*the origination of new species by natural causes*: a work which must soon bring about an entire revolution of opinion in favour of the grand principle of the self-evolving powers of nature.[4]

Darwin returned the favor, so to speak, by signing a letter, along with fellow scientists, countering the attack on *Essays and Reviews* made by Samuel Wilberforce, then bishop of Oxford, in *The Times*[5]—the same man who had harshly criticized Darwin's theories in the *Quarterly Review*[6] and in a debate at the annual meeting of the British Association for the Advancement of Science.[7]

As both a biblical scholar and an avid botanist, Hort filled his letters from 1860–61 with discussion of these debates. At one point his constant correspondent, Westcott, proposed that they and Lightfoot produce a series of mitigating essays representing "a mean between *Essays and Reviews* and Traditionalism."[8] Hort, for his part, proposed a declaration against "the suppression of free criticism" and "the adoption of a harsh and intolerant policy"[9] against the authors of *Essays and Reviews*. He espoused a position more sympathetic to the authors, writing to Westcott, "It is perhaps true that I feel the errors of the *E. and R.* less keenly than

---

4 Baden Powell, "On the Study of the Evidences of Christianity," in *Essays and Reviews*, ed. Shea and Whitla, 258. Emphasis original.

5 See Adrian J. Desmond and James Moore, *Darwin: The Life of a Tormented Evolutionist* (New York: Norton, 1991), 500–1.

6 Samuel Wilberforce, "On the Origin of Species," *Quarterly Review* 108 (1860): 225–64.

7 There are no verbatim records of the debate. See Keith Thomson, "Huxley, Wilberforce, and the Oxford Museum," *American Scientist* 88 (2000): 210–3 and Frank A. J. L. James, "An 'Open Clash between Science and the Church'? Wilberforce, Huxley, and Hooker on Darwin at the British Association, Oxford 1860," in *Science and Beliefs: From Natural Philosophy to Natural Science 1700–1900*, ed. David M. Knight and Matthew D. Eddy (Aldershot, UK: Ashgate, 2005), 171–94.

8 Brook Foss Westcott, *Life and Letters of Brooke Foss Westcott*, ed. Arthur Westcott, vol. 1 (London: Macmillan, 1903), 213–6, in letters to Hort, August 6, 1860; December 13, 1860; February 9, 1861, March 30, 1861; and to Lightfoot, December 20, 1860.

9 Hort, *Life and Letters*, 439, in a letter to Westcott, February 13, 1861.

you do. It appears to me tolerably certain that I have a stronger sense of their truths."[10]

Hort was also more accepting of Darwin's theories than was Westcott; when Westcott stated that he could find no evidence for Darwin's theories, Hort replied:

> It certainly startles me to find you saying that you have seen no facts which support such a view as Darwin's. But I do see immense difficulties in his theory, some of which might by this time have been removed, if he had understood more clearly the conditions of his problem, and made experiments accordingly. But it seems to me the most probable *manner* of development, and the reflexions suggested by his book drove me to the conclusion that some kind of development must be supposed.[11]

Hort's proposals for an article in *Macmillan Magazine* were along the same lines: he wished to clarify and organize Darwin's theories—to provide "a clear popular statement of the various facts of the argument"[12]—and not to write "about the 'moral' bearings of the subject: it is a ticklish matter."[13]

In fact, Hort understood the classification of species in a way quite similar to Darwin, viewing the Linnean system strictly as an artificial structure that does not necessarily convey actual underlying relationships. Hort says as much on two separate occasions, when recommending botany handbooks:

> Macgillivray[14] is perhaps the best for a beginning [in the study of botany], and you can easily show the boy that the Linnean system adopted bears little relation to the actual affinities in plants (which you must be able in a measure to detect), but it is an easy system for reference, and was in fact intended as no more by its great inventor.[15]

---

10 Ibid., in a letter to Westcott, February 15, 1861.

11 Ibid., 431, in a letter to Westcott, October 15, 1860. Emphasis original.

12 Ibid., 415, in a letter to Mr. A. Macmillan, March 10, 1860.

13 Ibid., in a letter to Mr. A. Macmillan, March 16, 1860.

14 William Macgillivray, *A Manual of Botany: Comprising Vegetable Anatomy and Physiology or The Structure and Function of Plants* (London: Scott, Webster, and Geary, 1840).

15 Hort, *Life and Letters*, 104, in a letter to John Ellerton, May 22, 1849.

To his mother he recommends John Lindley's *Ladies' Botany*[16] in part because it instructs the reader "not by telling you how to put plants in the shelves and compartments of any system invented by learned men, but helping you to see for yourself how they are actually related to each other in the unchangeable order of nature."[17]

Darwin writes in the same vein to the zoologist G. R. Waterhouse in 1843: "According to my opinion . . . classification consists in grouping beings according to their actual *relationship,* i.e. their consanguinity, or descent from common stocks."[18] This view is consistent with a more fully developed understanding of classification later published in *The Origin of Species*:

> Naturalists try to arrange the species, genera, and families in each class, on what is called the Natural System. But what is meant by this system? Some authors look at it merely as a scheme for arranging those living objects which are most alike, and for separating those which are most unlike; or as an artificial means for enunciating, as briefly as possible, general propositions,—that is, by one sentence to give the characters common to all carnivora, by another those common to the dog-genus, and then by adding a single sentence, a full description is given of each kind of dog. The ingenuity and utility of this system are indisputable. But many naturalists think that something more is meant by the Natural System; they believe that it reveals the plan of the Creator; but unless it be specified whether order in time or space, or what else is meant by the plan of the Creator, it seems to me that nothing is thus added to our knowledge. Such expressions as that famous one of Linnaeus, and which we often meet with in a more or less concealed form, that the characters do not make the genus, but that the genus gives the characters, seem to imply that something more is included in our classification, than mere resemblance. I believe that something more is included; and that propinquity of descent,—the only known cause of the similarity

16 John Lindley, *Ladies' Botany or A Familiar Introduction to the Study of the Natural System of Botany* (London: Ridgeway, 1834).

17 Hort, *Life and Letters,* 191, in a letter to his mother, March 8, 1851.

18 Charles Darwin, *Charles Darwin's Letters: A Selection, 1825–1859,* ed. Frederick Burkhardt (Cambridge: Cambridge University Press, 1996), 76, in a letter to G. R. Waterhouse, July 26, 1843. Emphasis original.

> of organic beings,—is the bond, hidden as it is by various degrees of modification, which is partially revealed to us by our classification.[19]

Darwin and Hort agree on the practicality of the Linnaean system, the ease with which one might use it to discuss groupings and similarities, but they also both emphasize that classification is merely the expression of something deeper, that is, the relationship between the varieties. The great difference between the two, at least in their quotations presented here, lies in Hort's belief in the "unchangeable order of nature" as shown through the relatedness of different species. Thus, for Hort, the structure of classification reveals (or should reveal) relatedness, which, in turn, reveals a greater, perhaps divine, order. For Darwin, however, the underlying "order" is nothing more than the branches of relations between varieties linked by consanguinity and descent—and it is certainly not "unchangeable":

> All the foregoing rules and aids and difficulties in classification are explained [above], if I do not greatly deceive myself, on the view that the natural system is founded on descent with modification; that the characters which naturalists consider as showing true affinity between any two or more species, are those which have been inherited from common parents, and, in so far, all true classification is genealogical; that community of descent is the hidden bond which naturalists have been unconsciously seeking, and not some unknown plan of creation, or the enunciation of general propositions, and the mere putting together and separating objects more and less alike.[20]

It must be remembered that Hort wrote of an "unchangeable order of nature" nine years before he read *The Origins of Species* and when he did so his first letter to Westcott on the topic was enthusiastically in its favor: "Have you read Darwin? How I should like a talk with you about it! In spite of difficulties, I am inclined to think it unanswerable. In any case it is a treat to read such a book."[21] At least as far as his letters show, Hort came, more or less, to agree with Darwin as to the manner in which different species came to be, through change and development over time. He

19 Darwin, *Origin of Species*, 413–4.

20 Ibid., 420.

21 Hort, *Life and Letters*, 414, in a letter to Westcott, March 10, 1860.

brushed aside arguments that declared the theory of evolution invalid because it could not pinpoint and describe the beginnings of development:

> I do not see why the inconceivableness of a beginning is any argument against any theory of development. The contrary theory is simply a harsh and contradictory attempt to conceive a beginning. That we are in doubt about the early history of organic life arises not from an impotence of conception, but from the mere fact that we were not there to see what, if it were taking place now, we certainly could see.[22]

In the matter of manuscript transmission, of course, there is no question that readings have changed and developed over time—not only is this obvious in an analysis of any two manuscripts, but it is the very reason why textual criticism exists at all. Similar to his belief that the identification of actual relationships between plant species, not simply superficial classification, underlies the true study of botany, Hort understood the identification of relationships between readings as crucial to the task of recovering the original text:

> In their *prima facie* character documents present themselves as so many independent and rival texts of greater or lesser purity. But as a matter of fact they are not independent: by the nature of the case they are all fragments, usually casual and scattered fragments, of a genealogical tree of transmission, sometimes of vast extent and intricacy. The more exactly we are able to trace the chief ramifications of the tree, and to determine the places of the several documents among the branches, the more secure will be the foundations laid for a criticism capable of distinguishing the original text from its successive corruptions. It may be laid down then emphatically, as a second principle, that ALL TRUSTWORTHY RESTORATION OF CORRUPTED TEXTS IS FOUNDED ON THE STUDY OF THEIR HISTORY, that is, of the relations of descent or affinity which connect the several documents.[23]

---

22 Ibid., 430, in a letter to Westcott, October 15, 1860.

23 Brooke Foss Westcott and Fenton Anthony Hort, "Introduction" in *The New Testament in the Original Greek*, vol. 2 (Cambridge: Macmillan, 1882), 39–40.

Just as grouping species together based solely on similar physical characteristics may lead to a misunderstanding of their actual descent and genealogical relationships, so grouping texts only by the similarity of their readings may lead to false conclusions. Thus, an understanding of the history of their transmission is imperative:

> Let it be supposed that a treatise exists in ten MSS. If they are used without reference to genealogy by an editor having a general preference for documentary evidence, a reading found in nine of them will in most cases be taken before a rival reading found only in the tenth, which will naturally be regarded as a casual aberration. If the editor decides otherwise, he does so in reliance on his own judgement either as to the high probability of the reading or as to the high excellence of the MS. He may be right in either case, and in the latter case he is more likely to be right than not: but where an overwhelming preponderance of the only kind of documentary evidence recognised is so boldly disregarded, a wide door is opened for dangerous uncertainty.[24]

Therefore, even the textual critic who knows that readings must be weighed and not counted can still be led astray if genealogical history is ignored.

In this way Hort expounds a view of the classification of texts and its underlying significance similar to Darwin's classification of species, in which "all true classification is genealogical" and the "community of descent is the hidden bond" which textual critics seek and *not* "the mere putting together and separating objects more and less alike" or "some unknown plan of creation," which might more explicitly be named the *Textus Receptus*. Building upon the classification of Griesbach, and ultimately Bengel, and the genealogical principles established by Lachmann, Hort relates four separate groups of readings (Alexandrian, Western, Syrian, and Neutral) more or less genealogically. This is done through a consideration of "Genealogical Evidence," which is the identification of which texts contain a conflation of readings from two or more groups of texts (and, therefore, must be posterior to those two or more groups),[25] and "Internal Evidence of Groups," which compares groups in terms of their tendencies to smooth over or let stand difficult readings, contain transcriptional errors, and so on:

---

24 Ibid., 40.

25 See ibid., 46ff.

> The evidence obtained in this manner is Internal Evidence, not Genealogical. But the validity of the inferences depends on the genealogical principle that community of reading implies community of origin. If we find, for instance, in any group of documents a succession of readings exhibiting an exceptional purity of text, that is, readings which the fullest consideration of Internal Evidence pronounces to be right in opposition to formidable arrays of Documentary Evidence, the cause must be that, as far at least as these readings are concerned, some one exceptionally pure MS was the common ancestor of all the members of that group.[26]

It is important to note that when Hort states "community of reading implies community of origin" he does not mean geographical origin but simply textual ancestry. In terms of their classification of manuscripts, Westcott and Hort consider geography of secondary importance and their assignation of the four group names is only loosely based on geography—in fact, the title "Western" is given only in keeping with tradition:

> We shall call the β group "Western," an appellation which has for more than a century been applied to its leading members. It was given at a time when the patristic evidence was very imperfectly known, and its bearing ill understood. . . . The fitness [of the name] is more open to question since it has become evident that readings of this class were current in ancient times in the East as well as the West, and probably to a great extent originated there. On the whole we are disposed to suspect that the "Western" text took its rise in North-western Syria or Asia Minor, and that it was soon carried to Rome, and thence spread in different directions to North Africa and most of the countries of Europe. From North-western Syria it would easily pass through Palestine and Egypt to Ethiopia. But this is at present hardly more than a speculation; *nor do any critical results depend upon it.* Whatever may have been the home of the "Western" text, a change of designation would now cause more confusion than it would remove, and it remains true that the only continuous and approximately pure monuments of the "Western" texts now surviving have every right to the name.[27]

---

26 Ibid., 60–1.

27 Ibid., 108. Emphasis added.

The names "Syrian" and "Alexandrian" denote locales of origin but, again, these origins play a minor role in text-critical judgments of antiquity and purity. At most, they help to sketch a rough history of the possible transmission of various versions and corruptions and possibly to explain the characteristic style of a group. The "Alexandrian" group, for example, is a pre-Syrian group relatively free of Western corruption, and it is probably due to its locale that it should be "extensively Non-Western":

> That a purer text should be preserved at Alexandria than in any other church would not in itself be surprising. There, if anywhere, it was to be anticipated that, owing to the proximity of an exact grammatical school, a more than usual watchfulness over the transcription of the writings of the apostles and apostolic men would be suggested and kept alive.[28]

But it is a small group of early Non-Western texts that interests Westcott and Hort the most and to which they give the name "Neutral." Hort locates the Neutral text outside Alexandria in various locales where it "survived the early Western inundation" and did so in "comparative isolation" and so avoided "extinction."[29] The use of such language calls to mind the observations Darwin made of populations of species living in isolation, without the changes effected by immigrating varieties.[30] Hort, however, writes of geographical isolation and distribution only briefly, without much consideration of how these elements might affect textual genealogy and mixture.

The Darwinian Theory of Local Texts

It is B. H. Streeter who, in his theory of "local texts," develops a distinctly Darwinian approach to the study of texts and their geographical origin.[31]

---

28 Ibid., 127.

29 Ibid., 129, 130, 127, respectively. On "The neutral text and its preservation" see §177–80, p. 126–30.

30 See particularly Darwin, *Origin of Species*, chapters 11–12, "Geographical Distribution."

31 This association of Streeter's local text theory and Darwin's theory of natural selection is first made in David C. Parker, *An Introduction to New Testament Manuscripts and Their Texts* (Cambridge: Cambridge University Press, 2008), 169: "Is there already an analogy between evolutionary biology and textual scholarship in Streeter's theory of local texts? Is Streeter drawing on Darwin's observations, for example on the differences between animals and plants in different parts of the Galapagos Islands? He writes that 'it is the

By the time Streeter published *The Four Gospels: A Study of Origins* in 1924,[32] evolution and natural selection—after a brief rejection of Darwinism at the turn of the century—had become widely accepted in the scientific community and beyond.[33] Clearly, at the outset of explaining his theory of local texts, Streeter has science, and particularly the scientific study of development, in mind:

> Many of those who recognise that both textual criticism and the analysis of sources are an essential preliminary to a truly historical investigation are nevertheless inclined to recoil from the study of these problems [i.e. concerning the history of the four Gospels], fearing lest they become choked by the dust of multifarious detail. To such I would venture to suggest that whether a particular investigation is instinct with interest or fraught with tedium depends very much on the spirit in which it is approached. The problems discussed in the present volume have, if one cares to look at it in that light, much the same kind of intellectual appeal as the quest for the solution of a difficult acrostic or of a problem in chess. An even better analogy would be the science of Geology; for that is recognised as truly a science, though science from which, from the nature of the case, is compelled to dispense with the method of experiment and relies solely on observation. *Geology attempts to reconstruct the history of the past by a highly scientific method of observation.* Facts, over as wide a range as possible, are collected, sifted, and compared in order that hypotheses may be framed which will satisfactorily account for

circumstance that several of the islands possess their own species . . ., these species having the same general habits, occupying analogous situations, and obviously filling the same place in the natural economy of this archipelago, that strikes me with wonder' (C. Darwin, *Journal of Researches into the Natural History and Geology of the Countries Visited during the Voyage of HMS Beagle*, London, 2003 (repr. of 3rd edn, 1860, 396)). His subsequent discussion of the reasons why this should be, including the significance of the natural barriers between the islands, has an analogy with the discussion of the degree of movement probable between different Christian centres which has been stimulated by Streeter's theory. The precise analogy between Streeter and Darwin is found in Streeter's theory that repeated reproduction of the text in a single locality led to the development of characteristics unique to that text. This is even natural selection in the sense that Streeter is more likely to have been thinking of unintentional change than of change made by conscious human intervention."

32 Burnett Hillman Streeter, *The Four Gospels: A Study of Origins* (London: Macmillan, 1924).

33 For a history of the "eclipse of Darwinism" and its later resurgence, see Peter J. Bowler, *Evolution: The History of an Idea*, rev. ed. (Berkeley: University of California Press, 1989), 246–81.

> the observed phenomena. *The critical investigations pursued in this volume are of a precisely similar character.* And the student who enters upon these problems in the same spirit of scientific inquiry as he would if they were problems of Geology will find the method not without interest and the results well worth the trouble.[34]

Geology, as it is presented here, is not the study of the composition of rocks and minerals but rather of the development of the Earth and the evolutionary history of life. When Streeter writes that the "critical investigations pursued in this volume are of a precisely similar character" as those in the field of geology, he does not exaggerate, for the questions he pursues as he studies extant manuscripts are the same as those asked in studying the geologic record: Why did certain varieties become extinct? Why is there a sudden population growth of this variety? How did this new variety appear? What were the environmental factors that shifted development in a certain direction?

Streeter believes that textual critics must answer two questions in particular with regard to textual variation in order to determine which readings might yield the original text:

> (1) [The textual critic] must account for the great divergence between the types of text current in the second, third and fourth centuries. (2) He must explain the origin of the Byzantine standard text and the process by which it replaced the other types. (3) Finally, in the light of the conclusions reached on these two points, he must endeavour to determine which of these types of text, or what kind of combination of them, will represent most nearly the text of the Gospels as they left the hands of their several authors. The third problem is of course much the most important; but he cannot hope to solve it rightly unless he has first found a reasonably satisfactory solution to the other two.[35]

In other words, the textual critic must explain the great diversity of early texts and then their eventual extinction or conformity to a standard text-type, the Byzantine or Majority text.

---

34 Streeter, *Four Gospels*, 23. Emphasis added.

35 Ibid., 30–1.

Streeter suggests that the provenance of readings—their relative isolation or proximity to other locales with different readings—accounts for various unique, hybrid, and commonplace readings observable in extant manuscripts.[36] Whether a text arose in a greater or lesser center of Christianity should also be taken into account as part of the "*conditions* which originally gave rise to the existence of these local texts," and help to pinpoint "the exact nature of the process of *progressive correction* into conformity with the Byzantine standard text."[37]

The language Streeter uses here evokes Darwin's descriptions of the appearance of varying species, but nowhere does Streeter channel Darwin more than in his theory of *how* differing conditions give rise to various text-types (and in a veritable echo of Darwin, Streeter includes an appendix titled "The Origin of Various Readings").[38] The first consideration is given to the way texts become "standard" and peculiar to a certain location. Streeter posits that as scribes in one area copy and recopy manuscripts, variants, of course, creep in. If the variants are not glaring errors and simply present a different reading, they are likely to remain undetected and be recopied into new manuscripts later. Eventually, the locale develops its own unique text-type:

> Now as soon as there were numerous copies of a book in circulation in the same area, one copy would constantly be corrected by another, and thus within that area a general standard text would be preserved. But what we have to consider is that it is unlikely that the errors in the first copy of the Gospel of John, for example, which reached Rome would be the same as those in the first copy which came to Alexandria; and as each of these would become the parent of most other copies used in those respective cities, there would, from the very beginning, be some difference between the local texts of Rome and Alexandria.[39]

---

36 Streeter is not the first textual critic to emphasize provenance in the study of manuscripts. That honor should perhaps go to J. M. A. Scholz, who took especial care in not only listing manuscripts but also their locales. See Johann Martin Augustin Scholz, *Novum Testamentum Graece*, 2 vols. (Leipzig: Fleischer, 1830–36). On Scholz see Samuel Prideaux Tregelles, *An Account of the Printed Text of the Greek New Testament* (London: Bagster, 1854), 94–5 and Bruce M. Metzger, *Chapters in the History of New Testament Textual Criticism* (Grand Rapids, MI: Eerdmans, 1963), 16–7.

37 Streeter, *Four Gospels*, 32. Emphasis added.

38 Ibid., 565–71.

39 Ibid., 35.

Here Streeter hypothesizes the effect of two slightly different texts arriving at different cities, which alone would account for variation between two areas, just as two slightly different groups of the same species, arriving in two different locales, would almost inevitably produce two distinct groups of descendents. More interesting, however, is Streeter's suggestion that there would be mutual corrections made between manuscripts of a certain area so that, with the accumulation of corrections and changes, a text-type is born. It is possible, he notes, but only just, that a text might develop in the same way in two different areas:

> Where a change would effect an obvious grammatical improvement or tend to assimilate the text of one Gospel to another, the same alteration might easily be made in two different neighborhoods. But only rarely would any of the causes of corruption result in a coincident alteration of exactly the same kind along two different lines of textual transmission. On the contrary, corruption as a rule causes texts to become in the course of time more and more different.[40]

Texts developing in an isolated region would be most likely to accumulate unique variants, especially during the earliest centuries of Christianity when there may have been less sharing of texts between different churches. The continued survival of oral tradition, as well as a flexible attitude toward the sacredness of the text of the New Testament, would compound the variations found in texts from separate areas. Unwritten stories might be added to the text and an accidental omission, if it appeared in one of the first copies of the text to reach an area, would be nearly impossible to detect: "for at a period when the churches were relatively isolated, a passage once omitted from the earliest copy which reached a particular church would not for a long while, if ever, be replaced."[41] The omission would then become a permanent feature of the local text.

Isolation functions in much the same manner in Darwin's theory of natural selection. A species in an area that is cut off from other regions, for example, on an island, would develop uniformly in one direction:

40 Ibid., 37.

41 Ibid., 36.

> Isolation, also, is an important element in the process of natural selection. In a confined or isolated area, if not very large, the organic and inorganic conditions of life will generally be in a great degree uniform; so that natural selection will tend to modify all the individuals of a varying species throughout the area in the same manner in relation to the conditions. Intercrosses, also, with the individuals of the same species, which otherwise would have inhabited the surrounding and differently circumstanced districts, will be prevented.[42]

As for the phenomenon of material being added or omitted from a text in a certain locale, a comparable development can be found in Darwin's understanding of use and disuse of a physical feature as the cause for the retaining or gradual loss of it:

> these several [climatic] considerations have made me believe that the wingless condition of so many Madeira beetles is due to the action of natural selection, but combined probably with disuse. For during thousands of successive generations each individual beetle which flew least, either from its wings having been ever so little less perfectly developed or from indolent habit, will have had the best chance of surviving from not being blown out to sea; and on the other hand, those beetles which most readily took to flight will oftenest have been blown to sea and thus have been destroyed.
>
> The insects in Madeira which are not ground-feeders, and which . . . must habitually use their wings to gain their subsistence, have . . . their wings not at all reduced, but even enlarged. This is quite compatible with the action of natural selection. For when a new insect first arrived on the island, the tendency of natural selection to enlarge or to reduce the wings, would depend on whether a greater number of individuals were saved by successfully battling with the winds, or by giving up the attempt and rarely or never flying.[43]

---

42 Darwin, *Origin of Species*, 104.

43 Ibid., 136. This understanding of use and disuse is practically the same as Lamarck's. See Ernst Mayr, "Introduction," in Charles Darwin, *On the Origin of Species; A Facsimile of the First Edition* (Cambridge, MA: Harvard University Press, 1964), xxv: "Curiously few evolutionists have noted that in addition to natural selection, Darwin admits use and disuse as an important evolutionary mechanism. In this he is perfectly clear."

Of course, the disuse Streeter speaks of—the omission of a portion of text—would have been missing from the very start, at the time a text is first introduced to a region, which is not an exact parallel to Darwin's theory of disuse, which posits a gradual diminishment of a physical feature. But the two theories would be more similar in the case of an omission of a very small portion of text that could go unnoticed by a copyist and so vanish over time from the standard local text. Additions to the text, even large ones, also function more similarly to Darwin's theory if, as Streeter hypothesizes, a story from oral tradition is added and, because of a more tolerant attitude toward changes in the text, copyists accept it and change other texts accordingly or produce new ones with the addition included.

So a local text, like a variety of a species, acquires its own unique character as it develops in a region, adapting, so to speak, to its particular environment. But this assumes a relative degree of isolation from texts of other regions, and of course eventually there would be increased communication between different centers of Christianity. Through communication, texts from one area may migrate to another and introduce changes into the local text, especially if a newly introduced text comes from a larger see with greater ecclesial authority:

> But, along with a growing veneration for the text as that of inspired Scripture, there would come a tendency, whenever a new copy of the Gospels for official use in the public services was wanted, to lay more and more stress on the importance of having an accurate text. This would naturally result in the smaller churches obtaining new copies from the greater metropolitan sees, since these would be thought likely to possess a pure text.[44]

If the points of communication between centers are understood as areas of geographic overlap, the hypothetical scenario presented by Streeter is almost exactly that of Darwin's understanding of how varieties of one species develop and intermingle over a broad area covering several regions:

> if the area be large, its several districts will almost certainly present different conditions of life; and then if natural selection be modifying and improving species in the several districts, there will be

44 Streeter, *Four Gospels*, 38.

> intercrossing with the other individuals of the same species on the confines of each. And in this case the effects of intercrossing can hardly be counterbalanced by natural selection always tending to modify all individuals in each district in exactly the same manner to the conditions of each; for in a continuous area, the conditions will generally graduate away insensibly from one district to another.[45]

Darwin presents here two different forces at work, that of natural selection through varying conditions in different districts and that of the intercrossing of different varieties of the same species. The latter may retard or change the direction of development through natural selection, and, with a species that reproduces rapidly, a new variety may quickly appear upon the borders of different districts and spread to other areas.[46]

This is, more or less, how Streeter imagines the gradual standardization of texts as the texts of larger sees halt and change the development of texts in smaller locales toward their own readings. The later communication and mingling of texts between the greater centers of Christianity would also drive the gradual development of a standard text: "Thus the local texts of smaller churches would tend to become assimilated to those of the greater centres in their immediate neighbourhood. The next stage would be for the great churches to compare their texts and endeavour to reach a standard text which would be universally accepted."[47]

An even more apt explanation of this process in biological terms comes from Darwin's description of the immigration of a vigorous variety into an area populated by the same species that had up to that point been isolated:

> If the country were open on its border, new forms would certainly immigrate, and this also would seriously disturb the relations of some of the former inhabitants. Let it be remembered how powerful the influence of a single introduced tree or mammal has been shown to be. But in the case of an island, or of a country partly surrounded by barriers, into which new and better adapted forms could not freely enter, we should then have places in the economy

---

45 Darwin, *Origin of Species*, 103.

46 Ibid.

47 Streeter, *Four Gospels*, 38.

> of nature which would assuredly be better filled up, if some of the original inhabitants were in some manner modified; for, had the area been open to immigration, these same places would have been seized on by intruders.[48]

If the more authoritative text from a great Christian see is understood as a better adapted form entering a smaller region newly open to immigration, then its eventual triumph as the dominant reading in that area is almost a matter of course. This more vigorous variety, spreading from region to region, would become the standard everywhere and "others will have to be improved in a corresponding degree or they will be exterminated."[49]

And this is what Streeter proposes happened in the rise of the Byzantine text: the local texts of smaller areas were assimilated until the Byzantine text became universally dominant. Of course, the readings from other great Christian centers, such as Alexandria, would, during this process, be mixed in with the standard.[50] Eventually, however, after a gradual process of a few hundred years, even these remnants of "pre-Byzantine" texts would be "revised away."[51] The very few non-Byzantine texts that survived did so only in highly remote areas that remained closed to the infiltration of the dominant text. In describing these texts Streeter explicitly refers to the similar phenomenon in nature:

> It is also noticeable that D was written by an ill-educated scribe, and that the same thing applies to other important MSS. with a large non-Byzantine element, e.g. L, Δ, 28 and, still more conspicuously, Θ. This suggests that they were written in out-of-the-way places, where the Byzantine text had not yet penetrated or had only recently done so. Zoology presents us with an analogy; the last survivors of a species, once widely prevalent but now on the way to extinction, are found in remote and isolated spots.[52]

48 Darwin, *Origin of Species*, 81–2.

49 Ibid., 106.

50 Streeter, *Four Gospels*, 41.

51 Ibid.

52 Ibid., 42.

These minority texts have no chance of survival in the broader, open regions where the Byzantine text has taken over. Their numbers dwindle and, as Darwin states: "Rarity, as geology tells us, is the precursor to extinction."[53]

The extinction of a text does not necessarily mean its physical destruction, for non-Byzantine texts do, of course, survive to the present day. The extinction of a text here means that it is taken out of circulation, that copies are no longer made from it. In biological terms, its reproductive life has ended and its genealogical line is cut off. Its fossils are preserved in the geologic record, but it has no descendents in the accruing layers.

Stagnation: An "Interlude"

Curiously, for the field of textual criticism, success, like rarity, has signaled the threat of its extinction. As traditionally defined by eighteenth- and nineteenth-century scholars, success is the attainment of the original and ancestral text of the New Testament. When Westcott and Hort boldly titled their edition *The New Testament in the Original Greek*, they were pronouncing victory at the climactic high noon of New Testament textual criticism. Consequently, the work that followed would seem to belong to the twilight of the field.

Such, of course, was not their intent. And, indeed, in the decades immediately following the publication of Westcott and Hort's edition, the field seemed to be in no danger of waning. New manuscripts were coming to light, including the Codex Washingtonius, Codex Koridethi, and the Chester Beatty Papyri, to name just a few. New evidence meant new classifications of readings and new theories regarding the history of the transmission of the New Testament, of which Streeter's is a prime example.

It was through Streeter's theories and his examination of a group of texts, already identified by Kirsopp Lake as related, that constituted what many considered the greatest advance in New Testament textual criticism in the fifty years after Westcott and Hort. Streeter hypothesized that family Θ, established by Lake, was indeed a distinct text-type belonging to none of the groups postulated by Westcott and Hort. By identifying its particular readings in certain of Origen's works and tracking Origen's geographical whereabouts during the composition of these works, Streeter concluded the text-type of family Θ was the text Origen used while residing in Caesarea.[54]

53 Darwin, *Origin of Species*, 109.

54 Streeter, *Four Gospels*, 77–108.

The establishment of the Caesarean text by Streeter seemed to be the next great step in determining the history of the transmission of the New Testament. Frederic Kenyon, in the 1932 Schweich Lectures, proclaimed it a triumph:

> There are from time to time books which mark a definite turning-point in the studies to which they relate. Streeter's *Four Gospels* is such a work in connexion with the textual criticism of the New Testament. Utilizing the work of others over a period of many years past, and adding thereto valuable contributions of his own, he has made good a definite addition to our knowledge, and established a fresh starting-place for further progress. Henceforward the Caesarean text has an assured place in textual criticism.[55]

Not only did Streeter locate family Θ in Caesarea through a brilliant study of Origen's writings and travels but he also reached a conclusion that supported his theory of local texts. The Caesarean family Θ presented a text already influenced by Byzantine readings (the Syrian text-type, in Westcott-Hort parlance) but also sharing some similarities to readings in Alexandrian and Western texts. This hybridity reflects exactly the characteristics of Streeter's local text, which would have its own distinctions yet bear the traces of intermingling with neighboring species. Indeed, in his "Concluding Survey" of the Koridethi manuscript and its family, Streeter states that the location of family Θ in Caesarea provides considerable confirmation for his theory of local texts from the preceding chapters of *The Four Gospels*, especially given the geographic locations of the major text-types and their affinities:

> If we look at a map we see at once that the Churches whose early texts we have attempted to identify stand in a circle round the Eastern Mediterranean—Alexandria, Caesarea, Antioch, (Ephesus), Italy-Gaul, and Carthage. The remarkable thing is that the texts we have examined form, as it were, a graded series. Each member of the series has many readings peculiar to itself, but each is related to its next-door neighbour far more closely than to remoter members of the series. . . .

55 Frederic G. Kenyon, *Recent Developments in the Textual Criticism of the Greek Bible*, Schweich Lectures of the British Academy, 1932 (London: Oxford University Press, 1933), 47.

> Antecedently we should rather expect the text of any particular locality to be, up to a point, intermediate between those of the localities geographically contiguous with it on either side. But the exactness of correspondence between the geographical propinquity and the resemblance of text exceeds anything we should have anticipated. And this fact is, I feel, of some weight in confirming the general thesis propounded in these chapters.[56]

Conclusions in scholarship, however, rarely remain unquestioned, and, four years after *The Four Gospels* was published, Lake, along with Robert Blake and Silva New, presented evidence that cast some doubt on Streeter's conclusions.[57] It seemed that Origen's travels from Alexandria to Caesarea did not coincide with his use of Alexandrian and Caesarean texts quite as neatly as Streeter proposed.[58] This did not necessarily disprove the identification of Caesarea as the eventual locale of family Θ, but it suggested the possibility that the text family may have originated in Alexandria and been brought to Caesarea by Origen himself.[59]

What called Streeter's entire thesis into question, however, was a review by F. C. Burkitt published in 1925, which argued that the manuscripts comprising family Θ (Codex Koridethi, families 1 and 13, 28, 565, 700) might not form a coherent group at all: "My chief objection to speaking of the 'Caesarean text' is that this term gives apparent definiteness and consistence to a set of 'various readings' that remain to me obstinately disparate and amorphous."[60] The way in which Streeter's text-types seemed to form a ring around the Mediterranean also struck Burkitt as too perfect: "I confess to a sort of prejudice against [the importance of the Caesarean text], a feeling that the actual facts of the textual tradition are not so easily described in terms of history and geography."[61] Other

56 Streeter, *Four Gospels*, 106.

57 Kirsopp Lake, Robert Blake, and Silva New, "The Caesarean Text of the Gospel of Mark," *Harvard Theological Review* 21 (October 1928): 206–404.

58 See ibid., 259–70.

59 See Kenyon, *Recent Developments*, 48–9.

60 Francis C. Burkitt, review of *The Four Gospels, a Study of Origins* by B. H. Streeter, *Journal of Theological Studies* 26 (1925): 284.

61 Ibid.

objections to a neatly delineated "Caesarean text" would follow.[62] By 1945 Bruce Metzger would state, in a far cry from Kenyon's confident praise of Streeter's theory:

> it must be acknowledged that at present the Caesarean text is disintegrating. There still remain several families—such as family 1, family 13, and the Armenian and Georgian versions—each of which exhibits certain characteristic features. But it is no longer possible to gather all these several families and individual manuscripts under one *vinculum* such as the Caesarean text.[63]

Still, Streeter's work left a lasting impression. In the 1973 Hatch Memorial Lecture, Eldon Epp pointed out that the supposed establishment of the Caesarean text-type was still being used as an argument for definite progress in the field, despite the fact that it had largely been refuted.[64] The title of the lecture was "The Twentieth Century Interlude in New Testament Textual Criticism," and in it Epp argued that the "Caesarean text affair" was simply another indicator that methodology had not advanced significantly since Westcott and Hort.[65] Not only was the discovery of a new text-type doubtful, but the very existence of text-types before the fourth century had been called into question by Kurt Aland.[66]

This latter proposal by Aland, that text-types had not yet formed in the second and third centuries but there existed instead a wide variety of texts as represented by early papyri, would seem to provide a new alternative to

---

62 For a summary of the criticisms of a "Caesarean text" expressed by Burkitt and other scholars such as P. L. Hedley, R. V. G. Tasker, and, perhaps most importantly, Teófilo Ayuso, see Metzger, "The Caesarean Text of the Gospels," *Journal of Biblical Literature* 64 (December 1945): 457–89, updated as a chapter with the same title in Metzger, *Chapters*, 42–72.

63 Metzger, "Caesarean Text," 483.

64 Eldon J. Epp, "The Twentieth Century Interlude in New Testament Textual Criticism," the W. H. P. Hatch Memorial Lecture, Society of Biblical Literature, November 11, 1973, at the Palmer House, Chicago, Illinois, repr. in *Perspectives on New Testament Textual Criticism: Collected Essays, 1962–2004* (Atlanta: Society of Biblical Literature, 2005), 69.

65 Ibid., 68–9.

66 See Kurt Aland, "The Significance of the Papyri for Progress in New Testament Research," in *The Bible in Modern Scholarship*, ed. J. Phillip Hyatt (Nashville, TN: Abingdon, 1965), 334–7. This skepticism is reiterated thirty years later in Kurt Aland and Barbara Aland, *The Text of the New Testament: An Introduction to the Critical Editions and to the Theory and Practice of Modern Textual Criticism*, 2nd ed., trans. Erroll F. Rhodes (Grand Rapids, MI: Eerdmans, 1995), 64.

Westcott and Hort's version of textual events.[67] And yet, as Epp contends, scholars have come no closer to forming a theory and history of the transmission of the New Testament:

> Yet, all of our past discoveries and also these present suggestions, as far as they concern the methodology and early history of the text, still leave us largely in the position of Westcott-Hort and within a twentieth century interlude in New Testament textual criticism and, what is more, they leave us with the major and decisive tasks still ahead.[68]

Epp goes on to argue that progress has also been lacking in the production of popular and critical editions of the Greek New Testament and in the development of methods in evaluating different readings.[69] This, along with a surprising return of the defense of the *Textus Receptus*, indicates that textual critics have traveled no great distance since Westcott and Hort and that the twentieth century has indeed been an "interlude" for New Testament textual criticism.[70]

There were, of course, objections to Epp's evaluation of the field.[71] But in several subsequent articles with, it seems, increasingly gloomy titles, Epp continued to write of the lack of methodological progress and, what is more, a decrease in interest in the field among biblical scholars.[72] Epp's student, Larry Hurtado, provides a far more optimistic perspective at

67 Hort understood Western readings to exist possibly before the end of the second century, Alexandrian readings no later than the beginning of the third century (possibly earlier), and Neutral readings even earlier than that. See Westcott and Hort, "Introduction," 122, 131.

68 Epp, "Twentieth Century Interlude," 79.

69 Ibid., 61–5, 80–3, 83–5.

70 Ibid., 85.

71 See especially Kurt Aland, "The Twentieth Century Interlude in New Testament Textual Criticism," in *Text and Interpretation: Studies in the New Testament Presented to Matthew Black*, ed. Ernest Best and Robert McLachlan Wilson (Cambridge: Cambridge University Press, 1979), 1–14.

72 See Eldon J. Epp, "The Eclectic Method: Solution or Symptom?" *Harvard Theological Review* 69 (1976): 211–57, reprinted in *Perspectives*, 125–73; "New Testament Textual Criticism in America: A Requiem for a Discipline," *Journal of Biblical Literature* 98 (1979): 94–8, reprinted in *Perspectives*, 175–84; and "A Continuing Interlude in New Testament Textual Criticism?" *Harvard Theological Review* 73 (1980): 131–51, reprinted in *Perspectives*, 183–209.

the turn of the millennium, pointing to increased publication and conference session attendance as evidence of a resurgent interest in textual criticism.[73]

Nevertheless, certain textual critics felt the need to write defensively regarding their work, oftentimes contrasting the philological work of textual criticism with that of literary criticism or theory in an attempt to revalidate the traditional aims of "lower criticism." These comparisons place consistent emphasis on interpretative criticism—or even new, non-traditional methods of textual criticism—as following passing fads and fashions. For example, Siegfried Wenzel, a Middle English philologist, writes:

> Philology is an attitude of respect for the *datum* . . . What, then, about "New Philology"?. . . If "New Philology" means that philology is currently moving a little more to the front, that is all to the better; the movement vindicates the faithfulness of those who have made its cultivation their main task. But as I have defined [traditional philology], "love of the word" that seeks understanding is a lasting concern of the intellectual life and as such stands above the currents of fashion. This is not to denigrate the many —-isms that strut for a while . . . Yet respect for the facts, for the concrete realities of the text, is and must remain basic.[74]

In a similar vein, G. Thomas Tanselle complains that textual criticism, "one of the few scholarly fields that can be talked about in terms of millennia," has, in an effort to adapt to changing methods, been "infiltrated with the fashionable buzz-words of literary theory."[75] New Testament textual critic David Parker defends his field by claiming that its development, even with its traditional aims, will outlast passing trends:

> There are even those who view [textual criticism] as a peculiarly obsolete aspect of western imperialism in its guise of the

73 Larry Hurtado, "Beyond the Interlude? Developments and Directions in New Testament Textual Criticism," in *Studies in the Early Text of the Gospels and Acts: The Papers of the First Birmingham Colloquium on the Textual Criticism of the New Testament*, Texts and Studies 1, ed. David G. K. Taylor (Atlanta: Society of Biblical Literature, 1999), 26–48.

74 Siegfried Wenzel, "Reflections on (New) Philology," *Speculum* 65 (1990): 17–8.

75 G. Thomas Tanselle, "Textual Criticism at the Millennium," *Studies in Bibliography* 54 (2001): 1–2.

> historico-critical method, as a dead member of a moribund body. I hope to convince one or two such individuals that textual criticism is ahead of them and not behind them.
>
> But current fashions in New Testament studies are not as significant as certain other contemporary developments [in textual criticism].[76]

These defensive statements are in response to a growing perception that textual criticism is arcane and dry, uninteresting and unprofitable. Not all of this can be blamed on the changing "fashions" of literary theory, as Bart Ehrman admits:

> A lot of fault for this perception lies with my colleagues in the field, who in fact are the worst you'll find anywhere at explaining why it is that what they do *matters* for anything, for example, for exegesis or historical reconstruction, let alone for the study of NT theology, the history of doctrine, or the social world of early Christianity.[77]

Yet poor articulation of the significance of textual criticism is but one reason for the endangerment of the field. Having declared the attainment of the original text as the foremost desideratum of New Testament textual criticism and understood the field as based on "technical expertise and positive knowledge,"[78] textual critics seem to have painted themselves

76 David C. Parker, *The Living Text of the Gospels* (Cambridge: Cambridge University Press, 1997), xi.

77 Bart D. Ehrman, "Text and Transmission: The Historical Significance of the 'Altered' Text," in *Studies in the Textual Criticism of the New Testament*, New Testament Tools and Studies 33 (Leiden: Brill, 2006), 341.

78 This description comes from Jerome McGann, "Preface, 1992," in *A Critique of Modern Textual Criticism*, paperback ed. (Charlottesville, VA: University of Virginia Press, 1992), xxiii, who writes of shifting trends in literary and textual criticism: "Literary criticism, a key cultural monitor, registers such [cultural] conditions and changes, and literary criticism has recently undergone a slight but dramatic shift toward textual studies in the largest sense. As these studies have begun to reacquire a self-conscious historical grasp of their work, they have also begun to reassert their ancient claim to authority in the general field of literary scholarship and criticism. That claim was undermined, I believe, during the long period (approximately from Karl Lachmann to Fredson Bowers) when textual criticism made a religion of technical expertise and positive knowledge. Not, of course, that textual studies—or literary studies in general—can ever dispense with such expertise and knowledge. But critical method and even factive materials are historically located, and they comprise a process by which learning is continually being reapproached and reimagined."

into a corner once that technical expertise has successfully attained the positive knowledge it claimed was its goal.

But have New Testament textual critics actually reached the original text? Whether they have or not, it was effectively communicated that they had, not only through the publication of Westcott and Hort's edition under the title *The New Testament in the Original Greek* but also through statements such as Kurt Aland's regarding the United Bible Society and Nestle-Aland's editions of the Greek New Testament:

> the new "Standard Text" has passed the test of the early papyri and majuscules. It indeed corresponds to the early text. The discrepancies that the witnesses of the early text show are within the normal range of variation found in the transmission of the New Testament. At no time or place do we find here readings that demand a change in the "Standard Text." When the research conducted here in all its brevity and compactness can be presented in full, the detailed apparatus provided for each variant will convince even the last doubter. A hundred years after Westcott-Hort, the goal of an edition of the New Testament "in the original Greek" seems to be achieved.[79]

Claims of this type are exactly those that give the impression that textual criticism, already classed as the lowly "drain clearer," "handmaiden," and "Cinderella" of literary studies,[80] has outlived its usefulness.

At this juncture, a rescue and revival of the field would come in the form of two new approaches. The first, to be discussed in the next chapter, uses computer technology; the second approach is the aforementioned

---

79 Kurt Aland, "Der neue 'Standard Text' in seinem Verhältnis zu den frühen Papryi und Majuskeln," *New Testament Criticism: Its Significance for Exegesis: Essays in Honour of Bruce M. Metzger*, ed. Eldon J. Epp and Gordon D. Fee (Oxford: Clarendon Press, 1981), 275: "der neue 'Standard-Text' hat die Probe an den frühen Papyri und Majuskeln bestanden. Er entspricht in der Tat dem Text der Frühzeit. Was dessen Zeugen an Abweichungen von ihm bieten, liegt in der normalen Variationsbreite neutastamentlicher Überlieferung. Nirgendwann und nirgendwo finden wir hier Lesarten, die eine Änderung des 'Standard-Textes' geböten. Wenn die hier in aller Kürze und Gedrängtheit angestellten Untersuchungen einmal vollständig vorgelegt werden können, wird der jeder Variante beigegebene ausführliche Apparat auch den letzten Zweifler davon überzeugen. Hundert Jahre nach Westcott-Hort scheint das Ziel einer Ausgabe des NT 'in the original Greek' erreicht."

80 See, respectively, Parker, *New Testament Manuscripts*, 181; Wenzel, "Reflections," 17; J. Eugene Botha, "New Testament Textual Criticism Is Dead! Long Live New Testament Textual Criticism!" *Hervormde Teologiese Studies* 63 (2007): 561.

New Philology, the development of which can be traced back to Streeter and his theory of local texts.

Deciphering the Geologic Record of the "Living" Text

This chapter has already addressed the many parallels between Darwin's theory of natural selection and Streeter's theory of local texts. However, a profound difference remains between Streeter's (and his contemporaries') perspective and Darwin's: Darwin emphasized *progress* in development while the former understood the development of texts as *corruption*. Species evolved, in Darwin's understanding, to their betterment and toward more perfect adaptation for survival. Texts, on the other hand, were understood by textual critics to degenerate the more they drifted away from their ancestral archetype.

Although he shared this understanding with his fellow textual critics, Streeter set a precedent for investigating the development of the text as part of early Christian history and not simply as the history of the transmission of corruption. His goal remained the uncovering of the original text, but he traced texts back to an archetype by casting them in a historical narrative. For example, Streeter's discussion of the terms "corruption" and "degeneration" with regards to "neutral" (B) and "Western" readings reveals the emphasis he places on restoring different local texts before proceeding to the task of determining the archetype:[81]

> If, however, we use a word like "corruption" or "degeneration" in [connection to the process of infiltration of the B text by "Western" readings], we must be on our guard against an easy fallacy. Any departures in these MSS. from the B type, which are of the nature of grammatical and stylistic correction, are "corruptions" in the strictest sense of the term; that is to say, they are alterations, intentional or accidental, of what the original authors wrote. But departures from the B text which consist in the substitution of a reading found in ancient authorities belonging to the Western family are corruptions in a different sense. In so far as they are departures from the oldest form of the Egyptian text they are a degeneration *of that*

81 It should be noted that Streeter here is arguing that B and, to a lesser degree, ℵ are not representatives of a "neutral" text as Westcott and Hort proposed but of a primitive Alexandrian text. See Streeter, *Four Gospels*, 54–61.

> *particular textual tradition.* But the Western authorities represent a textual tradition of great antiquity belonging to a different locality, and it may well happen that they sometimes preserve a true reading which has been lost in the B text. This is a point on which clear thinking is extremely necessary. To say that the text of B is a purer representative of its type than ℵ is by no means the same thing as saying that it is a purer representative than the original text of the Gospels as the authors wrote it.[82]

Thus, ℵ, considered less pure than B by Westcott and Hort, may actually preserve, in its "Western" readings (what could be categorized as "corruptions" *of the B text-type*), certain readings that are truer to the "original" than those in B, and these must not be mistaken for a primitive Alexandrian text, which Streeter believes B represents (and not a "neutral" text). The layers of different texts from different locales must first be understood and peeled back slowly before revealing their original form, no matter how impatient one is to arrive at that goal:

> The reader may ejaculate that he is not interested in the primitive text of Alexandria, but only in what the original authors wrote. In textual criticism there are no short cuts; and, since local texts of the Gospels came into existence in the second century, it will not be till we have got back to these in their most primitive form that we have all the materials on which to base our judgement as to what the original authors wrote.[83]

For Streeter, reaching "what the original authors wrote" not only involved tracing the stemmata of a family tree but also imagining a history that would account for its various branches and connections. This history considers the locales of different church sees, large and small, communications between them, the movement of various church authorities, and changing attitudes toward the texts of the New Testament. In short, Streeter envisions a *story* of the early Christian movement to explain the appearance of different readings in different locations. His work is the forerunner of what Parker would call "narrative textual criticism," which

82 Ibid., 60–1. Emphasis original.

83 Ibid.

would flourish toward the end of the twentieth century and the beginning of the twenty-first.[84]

As in Streeter's theory of local texts, variant readings in narrative textual criticism are not treated simply as disparate elements used to reach the original text but as part of the fabric of a narrative. So Parker states, "[Narrative textual criticism] represents a move away from the traditional atomizing of texts into variant readings, towards a reconstitution of the fragments into an intelligible whole."[85] Such as it is, narrative textual criticism emphatically denies that uncovering the original text is the only goal of textual criticism. Indeed, since Epp's watershed article, "The Multivalence of the Term 'Original Text' in New Testament Textual Criticism," it has become de rigueur to question the very existence of such an entity or at least to place the term in scare quotes.[86]

This shift from pursuing a unified, authoritative, original text to multiple narratives told through variants and varying manuscripts as diverse strands of Christian history represents a major change in scholarship, and narrative textual criticism can be classed as "an approach [that is] not untouched by postmodernism."[87] It seems that the trends, fashions, and fads once denigrated by textual critics have proved less ephemeral than once thought, as Epp states in 2005, in a retrospective collection of his articles:

> It is clear to me that we [textual critics] in our diverse intellectual world—if at times pulled and dragged—have moved beyond the "certainties" of historicism and even beyond the comforts of modernism to a recognition that the notion of "multiple truths" is more than a novel fad or a passing fancy. I believe that we now must acknowledge that claims to a single interpretation are increasingly difficult to justify in our multicultural and pluralist world—both secular and religious. . . . In New Testament textual criticism this emergent reality places at risk not only "*the* original text" but even

---

84 David C. Parker, review of *The Orthodox Corruption of Scripture: The Effect of Early Christological Controversies on the Text of the New Testament* by Bart D. Ehrman, Journal of Theological Studies 45 (1994): 704.

85 Ibid.

86 Eldon J. Epp, "The Multivalence of the Term 'Original Text' in New Testament Textual Criticism," *Harvard Theological Review* 92 (1999): 245–81.

87 Parker, review of Ehrman, 704.

> the concept of the "original text" as commonly understood, and it also jeopardizes not only "a single original reading" in a variation unit, but also the notion of a single "correct" interpretation of a variant.[88]

Epp himself published one of the first monographs of narrative textual criticism, *The Theological Tendency of Codex Bezae Cantabrigiensis in Acts*, which investigates the possible "anti-Judaizing" (or at least "universalizing") bent of readings found in D, a primary representative of the "Western" text-type.[89] Bart Ehrman adopts a different approach by examining readings from different texts, rather than within a single manuscript, as evidence of Christological debates in the second and third centuries.[90] Other examples of narrative textual criticism include Larry Hurtado's examination of manuscripts as early Christian artifacts, Wayne Kannaday's identification of apologetic maneuvers in readings in the Gospels, and Parker's description of gospel variants as comprising a "living text," a tradition of differing interpretations.[91]

In these varying approaches to narrative textual criticism the legacy of Streeter's work is evident: retelling the history of the transmission of a text as a narrative describing its social, political, and theological significance has become a legitimate goal in textual criticism.[92] To go back

---

88 Eldon J. Epp, "Introduction: A Half-Century Adventure with New Testament Textual Criticism," in *Perspectives*, xxxv.

89 Id., *The Theological Tendency of Codex Bezae Cantabrigiensis in Acts* (Cambridge: Cambridge University Press, 1966).

90 Bart D. Ehrman, *The Orthodox Corruption of Scripture: The Effect of Early Christological Controversies on the Text of the New Testament* (Oxford: Oxford University Press, 1993).

91 Larry W. Hurtado, *The Earliest Christian Artifacts: Manuscripts and Christian Origins* (Grand Rapids, MI: Eerdmans, 2006); Wayne C. Kannaday, Apologetic Discourse and the Scribal Tradition: Evidence of the Influence of Apologetic Interests on the Text of the Canonical Gospels (Leiden: Brill, 2004); Parker, Living Text.

92 A couple of dissenting voices should be noted. After the publication of Epp, *Theological Tendency*, and Charles S. C. Williams, *Alterations to the Synoptic Gospels and Acts* (Oxford: Blackwell, 1951), an even earlier example of the narrative textual criticism method, the following objection was voiced by Ernest C. Colwell, "Hort Redivivus: A Plea and a Program," in *Studies in Methodology in Textual Criticism of the New Testament* (Leiden: Brill, 1969), 150: "I use the phrase 'reconstruct the history of the manuscript tradition' with the assumption that the purpose of the reconstruction is to aid in the restoration of the original wording of the text. Thus I deny that those have salvaged historical value who point out that every variant may tell us something about the history of the church. The current enthusiasm for manuscript variations as contributions to the history of theology has no

to Streeter's analogy of geology, narrative textual critics now analyze the geologic record of texts, studying variants preserved in different layers of history and valuing them for their own sake rather than only as a means of reaching a mythical ancestor. This also reflects a Darwinian approach to the past: the investigation emphasizes the progressive development and adaptation of texts to the changing conditions of political power, theological debates, and socio-material contexts.

And although narrative textual critics have not explicitly drawn comparisons between their approach and that of evolutionary biologists, their work invites such analogies because they insist that the numerous variants of the New Testament make it a "living text." This phrase is not used simply to mean that the text is inspirational or vivid but that it in fact *behaves like a living organism.* Epp, for example, stresses this point in championing the legitimacy of narrative textual criticism:

> Numerous variant readings arose [through intentional scribal changes], yet we should not miss noticing that this scribal activity is another vivid evidence that the New Testament text was *a living text subject to the vicissitudes of existence—a living, breathing organism reflecting and reacting to its social and theological environment* as it moved along in the stream of the vibrant Christian community of which it was a part.[93]

---

solid foundation. Granted that a number of variant readings have been theologically motivated, would any serious student of the history of theology turn to these as a major source? Is it not true on the contrary that we can be sure of theological motivation for a variant reading only when the history of theology of that manuscript's time and place is already well known?" A more recent appeal to the traditional goals of textual criticism is given by Barbara Aland, "Welche Rolle spielen Textkritik und Textgeschichte für Verständnis des Neuen Testaments? Frühe Leserperspektiven," *New Testament Studies* 52 (2006): 304: "Ist nicht Textkritik eine historischkritische Wissenschaft per definitionem, die feststellt, was ist, und die gerade nicht in ungebührlichem Maß das subjektive Verstehen oder gar Willkür bei ihrer Entscheidungsfindung einbringen darf? Zwar ist auf das sog. innere Kriterium der Textkritik keineswegs zu verzichten, aber wie oft sind Entscheidungen nach dem inneren Kriterium umkehrbar, eben weil subjektiv. Wir bemühen uns daher im Institut für neutestamentliche Textforschung das äußere Kriterium zu stärken, d.h. die Einsicht in den Wert der Zeugen für sich und im Konspekt aller Zeugen, objektiv und rational zu begründen. . . . Ziel ist, und dazu bekenne ich mich ausdrücklich, dem Urtext oder besser: dem Ausgangstext der Überlieferung so nahe wie möglich zu kommen. Nur so gewinnen wir die Freiheit, um nach Textgeschichte und Verstehensgeschichte fragen zu können."

93 Eldon J. Epp, "Textual Criticism in the Exegesis of the New Testament," in *Perspectives*, 478. Emphasis added.

What Epp describes here is an organism in the Darwinian sense in that it changes and adapts. This biologizing of the text differs markedly from that of the eighteenth and nineteenth centuries, when textual critics understood variants and texts to represent static entities leading back to an original and which were to be mechanically (*sine interpretatione*) classified and charted. Epp's creature is far more organic and vibrant, to the point that manuscripts are "living and dynamic, with individual 'personalities.'"[94]

Parker uses the phrase in the very title of his book, *The Living Text of the Gospels*, and his meaning by the phrase is much the same as Epp's. The transmission of the New Testament represents a tradition of free interpretation of its words, as evinced by different readings, omissions, and additions, and the text is "alive" in that it changes and develops:

> harmonisers used the evangelists with the same degree of freedom as Matthew and Luke had used Mark. They reword and reorder the material. That such harmonisations are found centuries after the compilation of the Gospels is incontrovertible evidence that the traditions continued *to live, that is, to grow*.[95]

Parker gives an example of this growth in his discussion of the sayings on marriage and divorce found in the Gospels, noting that "Passages which were the focus of contentious issues were prone to change" as they were interpreted in different socio-historical contexts.[96] Thus, "if we abandon the a priori of single original texts, then we find that the saying(s) of Jesus on this subject had, as one might say, *a life of their own*."[97]

This usage of "living" is quite different from the meaning given to it in various passages of the New Testament, especially in the two places where "living" is used to describe the "word of God," the title given to the New Testament itself centuries later. Hebrews 4:12 emphasizes the ability of the "living word of God" to penetrate and lay bare all creatures

94 Id., "The Oxyrhynchus New Testament Papryi: 'Not Without Honor Except in Their Hometown'?" in *Perspectives*, 748.

95 Parker, *Living Text*, 205. Emphasis added.

96 Ibid., 90.

97 Ibid., 94. Emphasis added.

to judgment.[98] "The living and enduring word of God" of 1 Peter 1:23 is characterized by its imperishability,[99] a meaning often attached to the term, especially in the phrase "the living God" in both Hebrew scriptures and the New Testament. In the widespread usage of the phrase "the living God" in scripture, the focus is definitely *not* on the malleability and progressive growth of God but rather on God's immortal power, in contrast to inert and lifeless idols. However, the primary meaning evoked in the phrase "living text" by textual critics and the term "living" when used in the titles of various paraphrase versions of the bible,[100] is its ability to change and develop. The term emphasizes the continued relevancy of the text because of its adaptability to different times and cultures.

Even traditional textual critics describe the New Testament as a "living text." Kurt and Barbara Aland, for example, state:

> Until the beginning of the fourth century the text of the New Testament developed freely. It was a "living text" in the Greek literary tradition, unlike the text of the Hebrew Old Testament, which was subject to strict controls because (in the oriental tradition) the consonantal text was holy. . . . Variant readings in the New Testament text which are not due to simple scribal error . . . may be explained by its character as a "living text." While it is true that from at least the third century the scribes tried to copy their exemplars faithfully to the letter, they also followed the meaning as they transcribed the text . . . , and this gave rise to variants.[101]

---

98 "Indeed, the word of God is living and active, sharper than any two-edged sword, piercing until it divides soul from spirit, joints from marrow; it is able to judge the thoughts and intentions of the heart." (NRSV) *Ζῶν γὰρ ὁ λόγος τοῦ θεοῦ καὶ ἐνεργὴς καὶ τομώτερος ὑπὲρ πᾶσαν μάχαιραν δίστομον καὶ διϊκνούμενος ἄχρι μερισμοῦ ψυχῆς καὶ πνεύματος, ἁρμῶν τε καὶ μυελῶν, καὶ κριτικὸς ἐνθυμήσεων καὶ ἐννοιῶν καρδίας·* (NA 28)

99 "You have been born anew, not of perishable but of imperishable seed, through the living and enduring word of God. For 'All flesh is like grass and all its glory like the flower of grass. The grass withers, and the flower falls, but the word of the Lord endures forever.' That word is the good news that was announced to you." (1 Peter 1:23–5 NRSV) *ἀναγεγεννημένοι οὐκ ἐκ σπορᾶς φθαρτῆς ἀλλὰ ἀφθάρτου διὰ λόγου ζῶντος θεοῦ καὶ μένοντος. διότι πᾶσα σὰρξ ὡς χόρτος καὶ πᾶσα δόξα αὐτῆς ὡς ἄνθος χόρτου· ἐξηράνθη ὁ χόρτος καὶ τὸ ἄνθος ἐξέπεσεν· τὸ δὲ ῥῆμα κυρίου μένει εἰς τὸν αἰῶνα. τοῦτο δέ ἐστιν τὸ ῥῆμα τὸ εὐαγγελισθὲν εἰς ὑμᾶς.* (NA28)

100 See Tyndale House Publishers, *The Living Bible, Paraphrased*, ed. Kenneth N. Taylor (Wheaton, IL: Tyndale House, 1971); id., *The Catholic Living Bible, Paraphrased: A Thought-for-Thought Translation*, ed. Kenneth N. Taylor (Wheaton, IL: Tyndale House, 1976); id., *New Living Translation* (Wheaton, IL: Tyndale House, 1996).

101 Aland and Aland, *Text of the New Testament*, 69.

In some ways, the phrase "living text" is an apologia on the part of textual critics to account for a text believed by many to be authoritative, stable, eternal, and, therefore, *un*changing. The indignation caused by Mill's thirty thousand variants in 1707 is still echoed today by defenders of the *Textus Receptus* as the most trustworthy text of the New Testament.[102] In view of this insistence on a secure and unchanging text, textual critics recast the existence of change and the appearance of instability in a positive light. Rather than a textual tradition bereft of authority because of its variation, the New Testament is described as dynamic, vibrant, and spirited.

This new and invigorating perspective has the potential, however, to verge into negative and harmful comparisons. If texts can be "living," then they can also be "dead," if they, by textual critical standards, do not exhibit signs of evolutionary adaptation and variability.[103] The Alands, in the quotation above, compare the text of the New Testament to that of the Hebrew bible, which they describe as "subject to strict controls because (in the oriental tradition) the consonantal text was holy." This description stays clear of calling the Hebrew bible a "dead" text, but it depicts the difference between a living and non-living text in racial terms (rather than simply as a difference in transmission history), which can lead to quite negative implications.[104] In the same volume, the Alands again emphasize racial difference as the demarcation between textual traditions:

> The Greeks did not share the view of the orientals, for whom the letter had a sanctity of its own. The Hebrew text of the Old

---

102 See Edward F. Hills, *The King James Version Defended! A Christian View of the New Testament Manuscripts* (Des Moines, IA: Christian Research, 1956); David Otis Fuller, ed., *True or False? The Westcott-Hort Textual Theory Examined* (Grand Rapids, MI: Grand Rapids International, 1973); Maurice A. Robinson and William G. Pierpont, eds., *The New Testament in the Original Greek According to the Byzantine/Majority Textform*, (Atlanta: Original Word, 1991); Harry A. Sturtz, *The Byzantine Text-Type and New Testament Textual Criticism* (Nashville, TN: Nelson, 1984).

103 David Parker is the only author who explicitly states that he does not wish to imply that there are "dead texts." See David C. Parker, "*Et Incarnatus Est*," in *Manuscripts, Texts, Theology: Collected Papers 1977–2007* (Berlin: de Gruyter, 2009), 313: "The implication is not that there is also such a thing as a dead text, but that the text is living, living because it continues to change."

104 This emphasis on racial difference rather than the different transmission histories of texts is pointed out by James R. Royse, *Scribal Habits in Early Greek New Testament Papyri*, New Testament Tools, Studies and Documents 36 (Leiden: Brill, 2008), 25, and I am indebted to his collection of quotations as evidence for this emphasis given in the same volume, 25–6.

> Testament, like the text of the Quran, is alike in all manuscripts (except for unintentional errors). For Greeks it was the message contained that was sacred.[105]

It is possible to read this quotation without necessarily calling to mind European philhellenism and romantic nationalism; the authors could simply be offering a neutral comparison between two different scribal traditions. In fact, the same comparison could be made in a way that depicts the Greek scribes as falling short of Jewish standards. For example, Kenyon states: "Greek scribes never attained the meticulous accuracy of the copyists of the Hebrew Scriptures."[106] Nevertheless, the constant association of a "living text" with the "free" transmission of the New Testament certainly appears to show "oriental" texts and "copyists" (not scribes, per Kenyon) in a negative light. Elsewhere Kurt Aland asserts:

> And even when the canonicity of the twenty-seven texts of the New Testament was fixed by the church from the fourth century on, the text of the NT still remains a "living text." Well into the latest period, the scribes *felt free* to make changes and never copied their exemplar *as slavishly* as it was *the law* in the oriental tradition—whether with the Hebrew Old Testament or the Arabic Koran.[107]

And Colin H. Roberts writes, in the same vein:

> Greeks and Romans were acquainted with sacred books, whether those of minority groups such as Orphics or Pythagoreans or, as in Rome, belonging to the state, but the physical object was not treated with the same veneration nor the text itself so scrupulously

---

105 Aland and Aland, *Text of the New Testament*, 291. Quoted in Royse, *Scribal Habits*, 25.

106 Frederic G. Kenyon, *Books and Readers in Ancient Greece and Rome* (Oxford: Clarendon Press, 1932), 72. Quoted in Royse, *Scribal Habits*, 25.

107 Aland, "Der neue 'Standard Text,'" 263: "Und selbst als die Kanonizität der 27 Schriften des NT vom 4. Jahrhundert ab kirchlich fixiert wird, bleibt der Text des NT immer noch ein 'lebender Text.' Bis in die spätesten Zeiten hinein haben sich die Schreiber zu Änderungen frei gefühlt und ihre Vorlage niemals so sklavisch kopiert, wie das in der orientalischen Überlieferung Gesetz war—sei es beim hebräischen Alten Testament oder beim arabischen Koran." Emphasis added. The German is quoted in Royse, *Scribal Habits*, 25 n. 87.

> protected as was the case with the Jewish Law. The strictest rules governed the handling, the reading and the copying of the Law.[108]

These quotations emphasize a strict adherence to *law* and a veneration of the written letter of the text in the "oriental" tradition as opposed to a value for the *meaning* and *message* of the text in the Greek tradition, and, thus, in the tradition of the New Testament scribes. The description of a "slavish," legalistic attitude in the Jewish tradition is, of course, all too familiar to New Testament scholars, and it seems all the more a false caricature when considering the murky history of scribal traditions. In the same volume as Roberts's quotation above, Shemaryahu Talmon writes in complete contradiction to the above quotations:

> The resulting contact [in the Hellenistic period] with the Greek world of letters had a decisive impact on the transmission of the Old Testament. Jewish scribes emulated Greek scribal techniques and terminology, and adopted their insistence on exactitude in handing down written records and literary works.[109]

Regardless of which scribes from which culture insisted upon strict accuracy (and at what point in history), New Testament textual critics have, of late, emphasized that the diversity of readings found in the first few centuries of the transmission of the text is evidence of a living tradition that reveals the primordial vibrancy of the earliest Christian message. Kannaday, for instance, insists that the *spirit* of the Gospels persists despite the variations found in the text:

> For [the scribes of the Gospels], it seems, a text that bore testimony to a living spirit could easily afford a measure of liquidity in transmission. In a sense each word mattered, yet in another sense no word mattered; only The Word mattered, only the *Logos*.[110]

---

108 Colin H. Roberts, "Books in the Graeco-Roman World and in the New Testament," in *The Cambridge History of the Bible: From the Beginnings to Jerome*, vol. 1, ed. Peter R. Ackroyd and Christopher F. Evans (Cambridge: Cambridge University Press, 1970), 49. Quoted in Royse, *Scribal Habits*, 26.

109 Shemaryahu Talmon, "The Old Testament Text," in *The Cambridge History of the Bible*, 166. Quoted in Royse, *Scribal Habits*, 26.

110 Kannaday, *Apologetic Discourse*, 249.

Parker also interprets diverse readings as witnesses to a spirit of free interpretation, which was a part of the Christian community from the very beginning:

> does this remarkably free text [Codex Bezae] preserve the earliest Christian attitude to the tradition? That is, does its freedom reveal that it preserves *the spirit of the primitive use of Jesus' words*, precisely because the letter has been altered? If that is the case, then the quest for a single authoritative text is in itself a distortion of the tradition.[111]

Parker's answer to these questions is a definitive "yes." In his perspective, the manuscripts and their variants do not attest an original tradition that needs to be uncovered. Rather, this diversity *is* the tradition: "the manuscripts are not only transmitters of the tradition. They *are* the tradition."[112] Put another way, the free transmission of the New Testament is as Spirit-inspired as its initial composition:

> For the heart of the matter is that the definitive text is not essential to Christianity, because the presence of the Spirit is not limited to the inspiration of the written word. We have already approached this from the point of view of a false distinction between Scripture and tradition. Examining it in the present context, one is struck by the fact that a belief in single authoritative texts accords to the Spirit a large role in the formation of Scripture, and almost none at all in the growth of the tradition. Once the distinction has been abolished in the way that we have attempted above, it is possible fully to acknowledge that the very life and whole life of the church is in the Spirit.[113]

The living, spirited tradition of the New Testament presented by Parker and Kannaday, however, is in stark contrast to the "oriental" traditions evoked by other textual critics. There is, on the one side, the free, spirit-filled, living tradition of the Hellenistic West, the Christians. On the other side is the slavish, non-living textual tradition of the East, the orientals, the Jews, and

---

111 Parker, *Living Text*, 202. Emphasis added.

112 Ibid., 209.

113 Ibid., 211.

the Arabs.[114] In this way, race, culture, and artistic expression are generalized and bound together in a manner reminiscent of the romantic nationalism and understanding of *Volksgeist* popular in post-Enlightenment Europe. Just as Herder, then Hegel and Heidegger insisted on a free, authentic "spirit" as the underlying consciousness and bond of the German people, so biblical scholars claim a free and living primordial tradition for early Christianity.[115] It is no coincidence that both European nationalists and biblical scholars claim ancient Greece as their true heritage.[116] As Shawn Kelley states in his description of the influence of Hegel on biblical scholars in *Racializing Jesus*:

> For Hegel, and for most nineteenth-century thinkers (particularly within Germany), the Greeks represented something fundamentally new in the history of the world. With the Greeks we have the birth of the West and the triumph of Western individualism and freedom over Oriental despotism and servility. The Greeks come to symbolize freedom, individualism, and intellectual vibrancy, while the Orientals come to symbolize despotism, collectivity, and intellectual ossification. [F. C.] Baur takes over this fundamental antithesis between the Western (free) Greeks and the non-Western (servile) Orientals and interjects it into the very heart of his analysis of emerging Christianity.[117]

In the endeavor to "recover" the true, originary, free spirit of either a national people or of authentic Christianity, whatever is fettered and inauthentic must be jettisoned:

---

114 I do not mean to imply that *all* textual critics describe the Jewish or Arabic tradition this way; in fact, many, for example, Parker and Kannaday, do not address the textual transmission of other sacred writings in comparison to that of the New Testament. However, the upholding of the New Testament textual tradition as a "living" tradition raises the question: what of textual traditions that differ in history and scribal habits? Are they, then, dead? And does strict attention to the accuracy of copying from an exemplar mean that free *interpretation* is also taboo? I cannot accept this to be the case. Even if Jewish scribal habits were meticulously constrained, surely rabbinic interpretations as found, for example, in the Babylonian Talmud bear witness to a tradition of free and creative interpretation. In that case, what is the use of calling one scribal tradition "living" and another less so?

115 See Shawn Kelley, *Racializing Jesus: Race, Ideology and the Formation of Modern Biblical Scholarship*, Biblical Limits (New York: Routledge, 2002).

116 Textual critics claim their own field originated with the Greeks. See Bruce M. Metzger and Bart D. Ehrman, *The Text of the New Testament: Its Transmission, Corruption, and Restoration*, 4th ed. (New York: Oxford University Press, 2005), 197: "Like so many disciplines that we take for granted in our Western culture, textual criticism originated among the Greeks."

117 Ibid., 6–7.

> This primordial ground is not immediately accessible because it is buried deep in the past, at the misty moment of Christian origins. Traces of this originary ground can be glimpsed beneath layers of inauthentic, derivative, domestic, and hopelessly impure texts. It is the scholar's job to purge the authentic moment of these (racial, ethnic, female) impurities to open up the possibility of authentic encounter and to make possible an authentic grounding for Western identity.[118]

Kelley describes this effort to recover authenticity in various aspects of New Testament biblical scholarship, for example in parable scholarship and in the quests for the historical Jesus.[119] However, he neglects to mention textual criticism, and traditional New Testament textual criticism of the eighteenth through twentieth centuries could be described precisely as determined to "purge the authentic moment" (the "original text") of "hopelessly impure texts." But even with narrative textual criticism, and New Testament textual criticism in general in the twentieth century to the present, there is still the emphasis on a "free" and primeval spirit authentic to the vibrant message of earliest Christianity, as shown above.

What is more, textual critics not only describe early Christian scribes as belonging to the free, Greek, and Western tradition, but they also understand the history of textual transmission as following the general pattern, accepted by historians of early Christianity, of the free and authentic devolving into the inauthentic, tame, and domestic. As Ernest Colwell states:

> When we approach the larger task of writing the history of the entire manuscript period, we must start with a new characterization of the process. *The story of the manuscript tradition of the New Testament is the story of progression from a relatively uncontrolled tradition to a rigorously controlled tradition.*[120]

The terminology used to describe texts of the uncontrolled period as versus those of the controlled reflect exactly the values of the "aesthetic

---

118 Shawn Kelley, "Race, Aesthetics, and Gospel Scholarship: Embracing and Subverting the Aesthetic Ideology," in *Prejudice and Christian Beginnings: Investigating Race, Gender, and Ethnicity in Early Christian Studies*, ed. Laura Nasrallah and Elisabeth Schüssler Fiorenza (Minneapolis, MN: Fortress Press, 2009), 200.

119 See Kelley, *Racializing Jesus*, chapter 6; id., "Race, Aesthetics, and Gospel Scholarship," 200–8.

120 Colwell, "Hort Redivivus," 164. Emphasis original.

ideology" that Kelley argues is shared by both biblical scholars and post-Enlightenment philosophers:

> The positive values and the negative values are both conceived of in explicitly aesthetic terms, with the positive (that is, immediacy, dynamism, authenticity, primordiality, organicity) made possible by the purging of the negative (that is, what is mediated, inauthentic, static, inorganic).[121]

Thus, according to Colwell, "The general nature of the text in the earliest period (to AD 300) has long been recognized as 'wild,' 'uncontrolled,' 'unedited.'"[122] So, also, does Hort characterize the environment in which the "wild" text of the Western text-type arose:

> In surveying a long succession of Western readings by the side of others, we seem to be in the presence of a *vigorous and popular ecclesiastical life*, little scrupulous as to the letter of venerated writings, or as to their permanent function in the future, in comparison with the supposed fitness for *immediate and obvious edification*.[123]

The Western text is furthermore described in organic terms, not only as "wild" but also capable of wanton breeding: "They are corrupted texts due to the uncontrolled nature of their propagation."[124] Of course, one man's "corruption" is another man's "living text," and the apocryphal insertions and loose paraphrases of these early texts have come to be valued as evidence of the dynamic life of the early church communities.

But as church power moved East, along with Christian libraries and scriptoria, the "free" and "wild" text of the West became the domesticated text of the East, called the Majority, the Byzantine, or the Syrian text. It is important to note that although control and standardization is introduced more systematically into the transmission of the text during this period,

121 Kelley, "Race, Aesthetics, and Gospel Scholarship," 196.

122 Colwell, "Hort Redivivus," 166. Of course, from a traditional text-critical point of view, this "wildness" was detrimental to maintaining the "purity" of the original text. Nevertheless, Colwell's and Hort's comments quoted here also present their positive read of early Christian communities as dynamic.

123 Westcott and Hort, "Introduction," 126. Partially quoted in Royse, *Scribal Habits*, 21. Emphasis added.

124 Colwell, "Hort Redivivus," 166.

the Byzantine text is not, therefore, valued by traditional textual critics as less corrupt. This is for two main reasons: first, it is too far removed from earlier recensions and too late a development in the history of transmission; second, its text, though standardized to a large extent, tends to harmonize readings and make them more palatable, thereby introducing corruptions.

And it is these features, the smoothing away of rough edges and the systematic control of its reproduction, which also makes the Byzantine text of less value to the narrative textual critic. The vibrant immediacy of the living, organic text has become mediated, static, and tame. Thus, the evolving and adapting local text of Streeter's work and, likewise, the living, breathing organism of narrative textual criticism finally fall under the control of an authoritative, controlling power. This change in the species of the text is deemed to be of utmost importance in the history of its transmission:

> From what has been said, it follows that the dividing line in the history of the text is the achievement of control. This means editions with sanctions. In the period prior to this achievement, manuscripts with the same ancestry differ widely. After this achievement, manuscripts with the same ancestry differ so little that they become almost identical.[125]

What evolutionary development the text would undergo through natural selection, so to speak, is curbed by artificial, domesticating sanctions imposed upon it during the period of imperial Christianity. In Darwinian understanding, changes or the prevention of changes in the species are then divorced from the natural life of the organism itself:

> One of the most remarkable features in our domesticated races is that we see in them adaptation, not indeed to the animal's or plant's own good, but to man's use or fancy.[126]

---

125 Colwell, "Hort Redivivus," 168.

126 Darwin, 49. It should be noted that Darwin understood species to be far more diverse when bred under domestication than in nature, which is the opposite of the textual critical view of natural and domestic development (see Darwin, chapters 1 and 2). Nevertheless, the emphasis in this discussion is on the artificiality of selection by humans as versus "natural" causes. Whether Byzantine control of textual production can really be said to be more "artificial" than forces at work in smaller church sees is doubtful. Both are the product of human intention and will.

Of course, the period of standardization was finite, and new phases of transmission followed, up to the very present. So Parker concludes, in *The Living Text of the Gospels*:

> There was the emergence of the codex. There was a period of free development. There was also a move towards standard texts. There were translations into new languages and new cultures. There was continued interest in ancient copies, so that early text forms reappear centuries later. . . . We have seen how much the printed text has changed since 1516, and have wondered what changes will be brought by the electronic text. Today's influences on the text of the Gospels are not those of the early centuries, but they may prove equally powerful. The text of the Gospels remains a living text.[127]

It appears that so long as diversity and change are a part of the transmission of the New Testament, it remains a living text with organic development. In this perspective, the task of the textual critic is to understand its adaptation, endangerment, and growth. And just as the unifying forces of standardization squelched the authentic vibrancy of the tradition, so, too, does the insistence upon a single and authoritative original text.

But the flourishing of an evolutionary understanding of textual transmission has by no means signaled the end of traditional textual criticism. For, along with the work of Darwin, another nineteenth-century scientist would affect the direction of scientific and textual research in the twentieth century up to the present. Mendel's theory of inheritance and genetics would inspire new methodologies for unraveling the tangled ancestries of New Testament texts and for mapping its textual genome.

127 Parker, *Living Text*, 212–3.

4

Philology and Phylogeny

In fact it is hard to find a problem in the analysis of manuscript relationships that is not encountered in a different guise by a geneticist.

DAVID C. PARKER, "Through a Screen Darkly"

So similar are the techniques and difficulties in DNA evolution and literary text evolution, that each can be used to illustrate the other.

RICHARD DAWKINS, *The Ancestor's Tale*

Introduction: The Tree of Life

THE TREE OF Life stands at the beginning and end of the Christian bible, forbidden in Eden and freely available in the New Jerusalem. Its fruit offers immortality while its leaves bring healing, eradicating death and decay, the marks of time. The Tree of Life in science today is radically different. Instead of representing the absence of time, it illustrates the ravages of time and, ironically, the development and evolution of all life on Earth through destruction, disaster, and death.[1]

Indeed, one striking feature of the Tree of Life, as Darwin imagined it, is that most of it is dead. Only at its tips is it alive:

1 The two epigraphs come from David C. Parker, "Through a Screen Darkly: Digital Texts and the New Testament," *Journal for the Study of the New Testament* 25 (2003): 410; and Richard Dawkins, *The Ancestor's Tale: A Pilgrimage to the Dawn of Evolution* (New York: Houghton Mifflin, 2004), 128. Phylogenetics will be discussed later, but here I must pause to state that the phylogenetic Tree of Life does not actually illustrate *how much* time passes between the branches of the tree, but the very process of branching shows the effects of the passing of millions upon millions of years.

> As buds give rise by growth to fresh buds, and these, if vigorous, branch out and overtop on all sides many a feebler branch, so by generation I believe it has been with the great Tree of Life, which fills with its dead and broken branches the crust of the earth, and covers the surface with its ever branching and beautiful ramifications.[2]

But regardless of whether the branches of this Tree of Life are dead or alive, the most important idea they illustrate is the relatedness of all life, past and present. In Darwin's time the concept of common descent was virtually unheard of and even morally offensive in an age when the Great Chain of Being, though a concept past its prime, still held sway.[3] Furthermore, the only evidence Darwin could produce to support his preposterous theory was limited to contemporary knowledge of biological heredity—that, and reasoning by analogy:

> Analogy would lead me one step further, namely, to the belief that all animals and plants have descended from some one prototype. But analogy may be a deceitful guide. Nevertheless all living things have much in common, in their chemical compositions, their germinal vesicles, their cellular structure, and their laws of growth and reproduction. . . . Therefore I should infer from analogy that probably all organic beings which have ever lived on this earth have descended from some one primordial form, into which life was first breathed.[4]

Although Darwin and Mendel were contemporaries, Darwin seems never to have read Mendel's work (while Mendel certainly did read Darwin's *Origin of Species*).[5] Even if Darwin were aware of Mendel's laws of heredity, it is not a matter of course that he would have understood their relevance to his own theories. Those who did know of Mendel's work during his lifetime—his paper, "Versuche über Pflanzen-Hybriden," was read

---

2 Charles Darwin, *On the Origin of Species by Means of Natural Selection* (London: Murray, 1859), 130.

3 The best history of the Great Chain of Being remains Arthur O. Lovejoy's *The Great Chain of Being: A Study of the History of an Idea* (Cambridge, MA: Harvard University Press, 1963).

4 Darwin, *Origin of Species*, 484.

5 David Galton, "Did Darwin Read Mendel?" *Quarterly Journal of Medicine* 102 (2009): 587–9.

before the Society for the Study of Natural Science of Brunn in 1865, and forty offprints were sent out to prominent scientists[6]—did not grasp its importance.[7]

So it came about that Darwin's theory of evolution had no underlying, consistent theory of heredity to bolster its inference that all life on Earth is related. Only after the turn of the century did Mendelian genetics make an impact on evolutionary biology. To understand how modern genetics influences textual criticism, it will be useful to summarize Mendel's findings as well as the eventual synthesis of Darwinian evolution and Mendelian genetics. The most important concepts in Mendel's work are:

1. Heredity works through units and not blending. For example, if a brown-eyed woman and a blue-eyed man produced a son, his eye color would not be a blend of blue and brown (e.g., hazel) but either blue or brown.
2. Each hereditary unit (now called a "gene") determines one characteristic or trait.
3. Genes are always paired so that every organism has two copies of a gene.
4. A gene in a pair can either be dominant (A) or recessive (a).
5. Whenever a dominant gene exists in a pair (AA or Aa), the dominant trait is expressed; recessive traits are expressed only if both genes in a pair are recessive (aa).
6. The last point entails that the content of a gene pair and the expression of the pair in the organism are two different things. Today, the former is called the genotype and the latter the phenotype.
7. Gametes (sperm and egg cells) contain only one copy of each gene so that during sexual reproduction two separate sets of genes combine to form the full set of genes of a new organism. This, in essence, is

6 Ibid.

7 There are only eleven references to Mendel's "Versuche" before 1900, none of which shows an appreciation of its significance for heredity or evolution. See Robert Olby and Peter Gautrey, "Eleven References to Mendel before 1900," *Annals of Science* 24 (1968): 7–20. It is only in 1899, when Carl Correns first discussed Darwin and Mendel together in regard to hybridization, that the tide was about to turn (Correns would be one of three botanists who "rediscovered" Mendel in the twentieth century). Hence Olby and Gautrey conclude: "The moral to this account, it seems, is that one mention in the correct context in a popular and definitive treatise is worth any number of reports which are either out of their context or in such obscure or evanescent literature that no one would think of delving into them" (18).

Mendel's first law of inheritance, or the law of segregation: pairs of genes separate in cell division for the formation of a gamete.

8. The second law of inheritance is the law of independent assortment: pairs of genes separate and recombine independently of other pairs. For example, genes determining eye color separate and recombine independently of those for hair color.[8]

As these concepts became more widely known and understood after the turn of the century, the stage was set for the development of genetics, the modern synthesis of Darwinian evolution and Mendelian genetics, and phylogenetics, the study of evolutionary relationships between organisms on a molecular level.

In genetics, scientists discovered the location of genes on chromosomes, observed the division of chromosomes during meiosis (the division of cells for the production of gametes), and uncovered the structure of genetic material as double-helix strands of DNA (deoxyribonucleic acid), which are sequences of nucleotides containing adenine (A), thymine (T), guanine (G), or cytosine (C).

Eventually, the sequences of these nucleotides—making up the genetic code of an organism—could be mapped so that the location of specific genes on a chromosome could be identified. This gave rise to genomics, the mapping and study of the entire genetic sequence of an organism. With advances in genomics, scientists are now able not only to map entire genomes but also to use that information to determine genetic relationships between individual organisms, including, of course, humans.[9]

---

8 For a narrative account of Mendel's life and work, see Robin Marantz Henig, *A Monk and Two Peas: The Story of Gregor Mendel and the Discovery of Genetics* (London: Phoenix, 2001). For a discussion of Mendel's impact on modern science, see Peter J. Bowler, *The Mendelian Revolution: The Emergence of Hereditarian Concepts in Modern Science and Society* (London: Athlone, 1989).

9 For a good overall primer on genetics and the history of the field, see Barbara Wexler, *Genetics and Genetic Engineering*, Information Plus Reference Series (Detroit: ThompsonGale, 2011). Primary sources from major figures in the field include: Thomas Hunt Morgan, Hermann Joseph Muller, and Alfred Henry Sturtevant, *The Mechanism of Mendelian Heredity* (New York: Holt, 1926), which details the authors' findings regarding the location of genes on chromosomes; Alfred H. Sturtevant, *A History of Genetics* (New York: Harper & Row, 1965), which summarizes the history of genetics including the author's chromosomal mapping of the fruit fly; James D. Watson, *The Double Helix: A Personal Account of the Structure of DNA* (New York: New American Library, 1969); Eric S. Lander, Lauren M. Linton, Bruce Birren, et al., "Initial Sequencing and Analysis of the Human Genome," *Nature* 409 (2001): 860–921.

The first step toward the modern synthesis of Darwin and Mendel's ideas was taken when three scholars working independently proved through mathematical equation that, in a hypothetical environment without evolutionary forces at work, genotype frequencies remain stable and constant. That is to say, given equal numbers of pure dominants (AA), heterozygotes (Aa), and pure recessives (aa) that mate randomly, the dominant genotype does not take over a population but, rather, the ratio of all three types stays constant.[10] Of course, virtually no environment exists without natural selection or other forces at work, but the equation illustrating this equilibrium allowed scientists to measure quantitatively the effects of evolutionary forces in a given population.

All genes, and their variations ("alleles"), within a population came to be called a "gene pool," and scientists determined that the frequency of an allele within a gene pool could be affected through natural selection and genetic drift (random elimination or isolation of individuals).[11] Drastic changes within a gene pool could cause the emergence of a new species.[12] Together, Darwin's theory of natural selection and Mendel's principles of inheritance formed a new, dynamic way to understand biological variation, change, and history.

These advances in evolutionary biology, genetics, and genomics also gave scientists a new way to understand relationships between species and a method of taxonomy radically different from the Linnaean system. Rather than categorizing species by physical similarities alone (e.g., all species within the class *Mammalia* have females with mammary glands),

---

10 This is known as the Hardy-Weinberg equilibrium or (less frequently) the Hardy-Weinberg-Castle equilibrium, named after British mathematician Godfrey H. Hardy (1877–1947), German geneticist Wilhelm Weinberg (1862–1937), and American geneticist William E. Castle (1867–1962). See Godfrey H. Hardy, "Mendelian Proportions in a Mixed Population," *Science* 28 (1908): 49–50.

11 See John, B. S. Haldane, *A Mathematical Theory of Natural and Artificial Selection* (Cambridge: Cambridge University Press, 1924), which uses mathematical models to show the impact of natural selection on a population; Ronald A. Fisher, *The Genetical Theory of Natural Selection* (Oxford: Clarendon Press, 1930), which shows how the aggregate effect of different genes results in the variance observable in populations, demonstrating that Mendel is not at odds with Darwin; Sewall Wright, *Evolution and the Genetics of Populations*, 4 vols. (Chicago: University of Chicago Press, 1968–1978), which includes Wright's theory of genetic drift and fitness peaks; Theodosius Dobzhansky, *Genetics and the Origin of Species* (New York: Columbia University Press, 1937), which explains the mathematics at work behind population genetics and the evolutionary effects of changes within a gene pool.

12 See Ernst Mayr, *Systematics and the Origin of Species* (New York: Columbia University Press, 1942).

scientists can group animals according to their evolutionary relationships, based on shared ancestry. This taxonomy is called phylogenetic systemmatics or cladistics, and its results sometimes contradict traditional Linnaen classification in surprising ways.[13] For example, in cladistics, birds and crocodiles are grouped together as archosaurs—they share a common archosaur ancestor that branched off from the Diapsid line—while a Komodo dragon is left out of that group because it is a lepidosaur, even though it might seem to share far more observable similarities with the crocodile than the bird. In the same way, a zebra and a lungfish are more closely related than a trout and a lungfish, according to cladistics.[14]

These surprising groupings come about because, as its name implies, phylogenetics (*phylon*, Greek for "tribe" or "clan" and *genetikos/genesis*, "origin" or "source") is concerned solely with relationships based on evolutionary history and attempts to eschew all subjective classification by mere anatomical similarities or artificial, hierarchical ranks (the "kingdom," "phylum," "class," "order," etc. of the Linnaean system). It maintains a nominalist perspective—as pithily stated by systematists Lecointre and Le Guyader: "abstract principles exist not in nature but only in our words for them."[15] Phylogenetics shuns a teleological view of evolution that depicts homo sapiens as the pinnacle of "progress." Instead, it seeks to lay out an objective map of relationships based on phylogeny alone: the phylogenetic Tree of Life.[16]

Thanks to genomics, the Tree of Life can also now illustrate relationships on a molecular level. Similarities and differences in DNA sequences between different organisms can be used to construct the branches of the Tree, with computer software processing and analyzing enormous amounts of data. Multiple trees, representing different possible configurations of relationships, can be generated by software and then examined

---

13 The term "phylogenetic systemmatics" was coined by Willi Hennig, a German entomologist considered the founder of phylogenetics. See Willi Hennig, *Grundzüge einer Theorie der phylogenetischen Systematik* (Berlin: Deutscher Zentralverlag, 1950) or id., *Phylogenetic Systematics*, trans. D. Dwight Davis and Rainer Zangerl (Urbana, IL: University of Illinois Press, 1966).

14 For an expanded explanation of these examples, and a good, in-depth introduction to cladistics, see Guillaume Lecointre and Hervé Le Guyader, *The Tree of Life: A Phylogenetic Classification*, trans. Karen McCoy (Cambridge: Harvard University Press, 2006).

15 Ibid., 17.

16 To explore the phylogenetic Tree of Life, from root to tips, and its myriad branches, see David R. Maddison and Katja-S. Schulz, eds., *The Tree of Life Web Project* (2007), accessed April 23, 2013, http://tolweb.org.

and labeled "weak" or "robust" representations of evolutionary history, depending on statistical analysis.[17]

Darwin famously began the task of sketching the Tree of Life by writing in his notebook, "I think" and drawing a simple branching figure.[18] The phylogenetic Tree of Life now, when graphically represented, is dazzling in its complexity and beauty.[19] Usually presented as circular in shape, with intricate lines like the iris of an eye, its branches become denser as they grow away from its root, the ancient and enigmatic Last Universal Common Ancestor, LUCA.[20]

The Tree of Texts

As previously discussed, B. H. Streeter's theory of "local texts" and subsequent narratives describing the history of textual transmission have obvious analogies to Darwin's theories of natural selection and species adaptation.[21] The merging of Darwin's theories with Mendel's in the form of population genetics only strengthens the parallels between biological and textual evolution. Texts within a certain area at a point in history can be considered as making up a "gene pool" as scribes copy and create new manuscripts using one, some, or all available texts within their reach.[22]

---

17 See Lecointre and Le Guyader, *Tree of Life*, 25–37. On "bootstrapping" as a statistical analysis of a tree's robustness, see Joseph Felsenstein, "Confidence Limits on Phylogenies: An Approach Using the Bootstrap," *Evolution* 39 (1985): 783–91.

18 The sketch, made in 1837 in Darwin's *First Notebook on Transmutation of Species*, is on display at the Museum of Natural History in Manhattan. See also a copy online at "File:Darwin_tree," *Wikimedia Commons*, accessed April 23, 2013, http://commons.wikimedia.org/wiki/File:Darwin_tree.png.

19 For one beautiful representation of a tree mapping the relationships of 33,000 flowering plant species, see Michigan News, "U-M Biologist Plays Key Role in Effort to Create First Comprehensive Tree of Life," *University of Michigan News Service*, May 21, 2012, accessed April 23, 2013, http://ns.umich.edu/new/books-list/20400-u-m-biologist-plays-key-role-in-effort-to-create-first-comprehensive-tree-of-life.

20 Also known as the Last Universal Ancestor (LUA), this organism is believed by some to be a protoeukaryote, possibly living more than three billion years ago. See Nicolas Glansdorff, Ying Xu, and Bernard Labedan, "The Last Universal Common Ancestor: Emergence, Constitution and Genetic Legacy of an Elusive Forerunner," *Biology Direct* 3 (2008), accessed April 24, 2013, http://www.biology-direct.com/content/3/1/29, especially the section "Palaeochemistry and chronology of early eukaryote evolution" and the bullet points under "Conclusion."

21 See Chapter 3.

22 That scribes could create a new manuscript from more than one copy is what makes the NT textual tradition so "contaminated." See Chapter 2.

"Genetic drift" can be compared to the effects of the destruction of a random selection of a group of manuscripts, for example by a fire in a library or by hungry mice in a monastery. Intentional destruction of certain texts due to their objectionable contents can be considered a force of natural textual selection, whether theological, political, or social.

But genetics also offers a completely new metaphor through which to view texts and textual variants, and the influence of genetics and genomics on textual criticism is apparent even on a cursory reading of textual critical scholarship. As early as 1963, Ernest Colwell and Ernest Tune advised that variants should be "classified as either genetically significant or not."[23] Variants subsequently have been understood in precisely this way, as influential or not in altering a textual gene pool and textual relationships.[24]

In many ways, the goals of phylogenetics and textual criticism are the same: to determine the evolutionary relationships between organisms/texts and, ultimately—at least in the case of traditional textual criticism and one aspect of phylogenetics—to identify and describe the Last Universal Common Ancestor. Even the debate over whether it is necessary or even possible to identify LUCA or the "original text" is shared by both fields. And the very meaning of LUCA and "original text" is also disputed. The Last Universal Common Ancestor does not indicate the earliest living organism, only the earliest organism to be ancestor to all subsequent life on earth. It may not even indicate one individual species but a community of organisms.[25]

Similarly, the textual critical terms "original text," "autograph," and "archetype" have become fraught with qualifications. Does "original"

23 Ernest C. Colwell and Ernest W. Tune, "Method in Establishing Quantitative Relationships between MS. Text-types," in *Biblical and Patristic Studies: In Memory of Robert Pierce Casey*, ed. J. Neville Birdsall and Robert W. Thomson (Freiburg: Herder, 1963), 59. The term "genetics" was coined by William Bateson in a letter to geologist Adam Sedgwick in 1905; he used it publically for the first time at the Third International Congress on Hybridization and Plant Breeding in 1906. See Ted Everson, *The Gene: A Historical Perspective* (Westport, CT: Greenwood, 2007), 49. The discovery of the structure of DNA as a double helix by Watson and Crick was made in 1953, and is considered the beginning of modern genetics.

24 This is not to imply that variants are *always* discussed using this metaphor, but the comparison is prevalent enough. See Eldon J. Epp, "Toward the Clarification of the Term 'Textual Variant,'" in *Perspectives on New Testament Textual Criticism: Collected Essays, 1962–2004* (Atlanta: Society of Biblical Literature, 2005), especially the section "Significant and Insignificant Readings," 116–21. Also, Parker, "Through a Screen Darkly," 401, on the creation of new editions by textual critics, who are, in a sense, simply creating more texts to join the masses: "what one is doing is readjusting the textual genome pool."

25 See Glansdorff, Xu, and Labedan, "Last Universal Common Ancestor"; also Carl Woese, "The Universal Ancestor," *Proceedings of the National Academy of Sciences* 95 (1998): 6854–9.

mean the first manuscript produced by the author's hand? Or is it the first copy the author (or others) deemed worthy for wider circulation?[26] One prevalent term now used in New Testament textual criticism is the *Ausgangstext* or the "initial text," that is, the text at the start of circulation, the start of its genetic evolution.[27]

Whatever the "original text" or *Ausgangstext* may be, the continuing search for origins, as alive today as in the nineteenth century, has fueled textual criticism's use of phylogenetics for tracing the genetic ancestry of texts. Because there are so many apparent parallels between an organism's genes and a manuscript's text, the tools of genomics and cladistics are readily adaptable to text-critical work. The most thoroughgoing use of phylogenetics in textual criticism is found in The Canterbury Tales Project, based at the University of Birmingham, which transcribes, collates, and maps the textual history of *The Canterbury Tales,* through computer analysis.[28]

In order to be processed by the software phylogenists use to create trees of organisms, a manuscript's text must first be transcribed, lined up with matching texts from other manuscripts of the same work, and collated. A code is generated from this comparison so that, for example, A = a baseline text (this does not indicate priority), B = word omitted, C = change of spelling affecting meaning, and so on (see Table 4.1). In this way, each line of text is coded and then analyzable by a program to produce a phylogenetic tree.[29]

The two programs most commonly used to analyze these codes and create a phylogeny of texts are Phylogenetic Analysis Using Parsimony

---

26 For other possibilities, see Eldon J. Epp, "The Multivalence of the Term 'Original Text' in New Testament Textual Criticism," *Harvard Theological Review* 92 (1999): 245–81.

27 See Gerd Mink, "Problems of a Highly Contaminated Tradition: The New Testament: Stemmata of Variants as a Source for a Genealogy of Witnesses," in *Studies in Stemmatology II,* ed. Pieter van Reenan, August den Hollander, and Margot van Mulken (Amsterdam: John Benjamins, 2004), 25: "The initial text is a hypothetical, reconstructed text, as it presumably existed, according to the hypothesis, before the beginning of its copying. . . . The initial text is not identical with the original, the text of the author. Between the autograph and the initial text considerable changes may have taken place which may not have left a single trace in the surviving textual tradition. Even if this is not the case, differences between the original and the initial text must be taken into account."

28 See *Canterbury Tales Project,* accessed July 22, 2015, http://hridigital.shef.ac.uk/canterbury-tales.

29 See Adrian C. Barbrook et al., "The Phylogeny of *The Canterbury Tales,*" *Nature* 394 (1998): 839; also Christopher J. Howe et al., "Manuscript Evolution," *Trends in Genetics* 17 (2001): 147–52.

Table 4.1. Adapted from Christopher J. Howe, Adrian C. Barbrook, Matthew Spencer, et al.,"Manuscript Evolution," *Trends in Genetics* 17 (2001), 148.

The transcription of a line from John Lydgate's *Kings of England* (15th c.) from seven sources:

ffourtene yeere he bare his crowne l reede	Ashmole 59
xiiij^{e} yere he bare his crowne in dede	Bodley 48
xiiij^{e} yere bare his corone in dede	Bodley 686
ffourtene yere he bare his croune I rede	CUL Ad6686
ffourtene yer^{e} bare he his crowne in dede	Harley 2261
fortene bare hys crown in dede	Landsdowne 210
Bare th^{e} crowne xij yere xi monthes & xvi dayes in dede	de Worde (printed)

The texts are aligned:

ffourtene	yeere	he	bare	his	crowne	I	reede
xiiij^{e}	yere	he	bare	his	crowne	in	dede
xiiij^{e}	yere		bare	his	corone	in	dede
ffourtene	yere	he	bare	his	croune	I	rede
ffourtene	yer^{e}	bare	he	his	crowne	in	dede
fortene			bare	hys	crown	in	dede
xi	yere xi monthes & xvi dayes		bare	th^{e}	crowne	in	dede

Codes are assigned to the types of variants in the texts and the lines are thus encoded:

AAAAAAAHH	Ashmole 59
AAAAAAAAA	Bodley 48
AAASAAAAA	Bodley 686
AAAAAAAHH	CUL Ad6686
AAARAAAAA	Harley 2261
ALASAAAAA	Lansdowne 210
HAESRMAAA	de Worde

(PAUP) and SplitsTree.[30] These programs create trees based on the degree of resemblance between texts. PAUP creates trees based on the rule of parsimony, that is, its trees explain the relationship of variations based on the fewest possible changes between texts (in other words, following

30 Peter Robinson, "New Methods of Editing, Exploring, and Reading *The Canterbury Tales*," based on a talk at the conference "I nuovi orizzonti della filologia," Accademia Nazionale dei Lincei, Rome, May 28, 1998, accessed July 22, 2015, http://www.petermwrobinson.me.uk/canterburytalesproject.com/pubs/desc2.html. For a downloadable copy of the

Ockham's razor).[31] SplitsTree is a useful program because it can create networks rather than bifurcating trees (trees with branches that divide into two), which is useful for working with manuscripts that have a complex phylogeny.[32] A bifurcating tree cannot adequately present the history of textual transmission if certain manuscripts are copied from more than one source or if a source gives rise to several different manuscripts at around the same time, and this is where SplitsTree has an advantage over PAUP.[33]

In biological evolution, a bifurcating tree is usually the structure representative of the emergence of a new species: rarely do several new species branch off from an ancestor all at once (this phenomenon is called a "hard polytomy").[34] But the emergence of several different texts from one exemplar can occur quite commonly in a manuscript tradition, and there are other aspects of textual evolution that seem at odds with biological evolution. Another significant difference is that in biological evolution, an ancestor species continues to evolve, so that the original ancestral form ceases to exist. Therefore, the "tips" of phylogenetic branches are

---

PAUP program and further information, see PAUP, Sinauer Associates, Inc., accessed April 25, 2013, http://paup.csit.fsu.edu/. For the SplitsTree program and information, see "SplitsTree4," *Algorithms in Bioinformatics*, Eberhard Karls Universität Tübingen, accessed April 25, 2013, http://www.splitstree.org/.

31 See Arthur R. Lee, "Numerical Taxonomy Revisited: John Griffith, Cladistic Analysis and St. Augustine's Quaestiones in Heptateuchum," in *Studia Patristica Vol. XX: Papers Presented at the Tenth International Conference on Patristic Studies Held in Oxford 1987*, ed. Elizabeth A. Livingstone (Leuven: Peeters, 1989), 24–32, on early uses of two similar programs (PHYLIP and MacClade) prior to the development of PAUP; for work with PAUP on the Old Norse Svipdagsmál cycle, see Peter Robinson and Robert O'Hara, "Cladistic Analysis of a Norse Manuscript Tradition," in *Research in Humanities Computing*, vol. 4, ed. Susan Hockey and Nancy Ide (Oxford: Oxford University Press, 1996), 115–37; for use of PAUP by The Canterbury Tales Project, see Peter Robinson, "A Stemmatic Analysis of the Fifteenth-Century Witnesses to The Wife of Bath's Prologue," in *The Canterbury Tales Project Occasional Papers*, vol. 2, ed. Norman Blake and Peter Robinson (Oxford: Office for Humanities Communication, 1997), 69–132.

32 See Howe et al., "Manuscript Evolution," 149; for more on SplitsTree, see Daniel H. Huson, "SplitsTree: Analyzing and Visualizing Evolutionary Data," *Bioinformatics* 14 (1998): 68–73. One method of generating networks of manuscripts with SplitsTree that has become quite useful is NeighborNet; see David Bryant and Vincent Moulton, "NeighborNet: An Agglomerative Method for the Construction of Planar Phylogenetic Networks," in *Algorithms in Bioinformatics: Second International Workshop, WABI*, ed. Roderic Guigó and Dan Gusfield, Lecture Notes in Computer Science 2452 (Berlin: Springer, 2002), 375–91.

33 Christopher Howe et al., "Parallels between Stemmatology and Phylogenetics," in *Studies in Stemmatology II*, ed. Pieter van Reenen, August den Hollander, and Margot van Mulken (Amsterdam: John Benjamins, 2004), 6.

34 Matthew Spencer et al., "Phylogenetics of Artificial Manuscripts," *Journal of Theoretical Biology* 227 (2004): 503–4.

contemporary species that do not include ancestral forms. But in manuscript transmission, of course, the text of an ancestral manuscript that is extant has not evolved continuously along with its textual progeny.[35]

Given these differences, the suitability of programs like SplitsTree for use with manuscripts might seem questionable. A group of scholars, including some from The Canterbury Tales Project, resolved to test the appropriateness of phylogenetic methods for textual criticism by simulating the transmission of a text: first, a handwritten copy of a printed text was created; from this copy were generated (whether directly from it or from one of its descendents) twenty more by hand using graduate students from the University of Cambridge as "scribes" who were unaware of the purpose of the study.[36] The "true phylogeny" was known to the conductors of the study but unknown to the biologist inputting the data of the different manuscripts into PAUP.[37] The most "robust" trees generated by the program were surprisingly close to the true phylogeny, demonstrating the usefulness, if not the perfect accuracy, of phylogenetic methods for textual criticism.[38]

So even though hard polytomies occur in manuscript transmission, and there is no continual evolution of ancestral texts, cladistics seems to work, at least on a broad scale, for creating textual trees. And for almost every problem encountered by textual critics in tracing a textual ancestry, there seems to exist an analogous phenomenon in biology. For example, the textual phenomenon of "contamination," that is, when a scribe creates a new text by copying from more than one exemplar,[39] is comparable to genetic recombination, which happens most frequently during meiosis (cell division for sexual reproduction) when two chromosomes come very close together for the process and, when they cross over, homologous segments of DNA are swapped between the two.[40]

---

35 Ibid., 504. A case could be made that an extant, ancestral manuscript could continue to evolve if it is corrected or used in a palimpsest.

36 Ibid.

37 Ibid., 505. Two different tree reconstruction methods in PAUP were used, and the bootstrap method for analyzing robustness was used.

38 Ibid., 509–10.

39 See Chapter 2.

40 See Howe et al., "Manuscript Evolution," 150, for the comparison of the two. For more on genetic recombination, see Bruce Alberts et al., "Genetic Recombination," in *Molecular Biology of the Cell*, 3rd ed. (New York: Garland Science, 1994), available from the National Center from Biotechnology Information, accessed April 29, 2013, http://www.ncbi.nlm.nih.gov/books/NBK28388/.

If, however, a scribe copies partly from one manuscript and partly from a manuscript of an entirely different work, that is more closely parallel to the lateral (or "horizontal") transfer of genes between two different organisms outside of sexual reproduction, between two different species (usually bacteria).[41] Because these are problems that biologists also face in genetic cladistics, new programs and adjustments to existing software are continually being developed to handle these complexities.[42] In fact, recent studies show that a new method, called maximum chi-squared analysis, can detect, in cases of contamination, when there is a change of exemplar in the copying of a text.[43] Test cases show that the results of using this analysis with phylogenetic programs can produce results in close agreement with stemmata created through traditional, philological methods, signifying that phylogenetic techniques may be equal to the task of creating reliable textual stemmata, even in the face of contamination.[44]

Another problem textual critics face that phylogenetic stemmatics solves is the circular reasoning that underlies traditional philological stemmata: in order to create a stemma of texts, an original, ancestral text must be chosen from which the stemma can be created—to determine the original text![45] But, with phylogenetics, an "unrooted tree" can first be created, which has no predetermined ancestral root and no chronology implicit in its network of textual relationships. Thus, a textual critic can

---

41 Howe et al., "Manuscript Evolution," 150. For more on the lateral transfer of genetic material, see Howard Ochman, Jeffrey G. Lawrence, and Eduardo A. Groisman, "Lateral Gene Transfer and the Nature of Bacterial Innovation," *Nature* 405 (2000): 299–304.

42 For examples of innovations used to account for lateral gene transfer in phylogenetics, see John Maynard Smith, "Analyzing the Mosaic Structure of Genes," *Journal of Molecular Evolution* 34 (1992): 126–9, which presents maximum chi-squared analysis, and Robert G. Beiko and Nicholas Hamilton, "Phylogenetic Identification of Lateral Genetic Transfer," *BioMedical Central Evolutionary Biology* 6 (2006), accessed July 27, 2015, http://www.biomedcentral.com/content/pdf/1471-2148-6-15.pdf.

43 See Heather F. Windram, Prue Shaw, Peter Robinson, and Christopher J. Howe, "Dante's *Monarchia* as a Test Case for the Use of Phylogenetic Methods in Stemmatic Analysis," *Literary and Linguistic Computing* 23 (2008): 443–63 and Heather F. Windram, Christopher J. Howe, and Matthew Spencer, "The Identification in Exemplar Change in the *Wife of Bath's Prologue* Using the Maximum Chi-Squared Method," *Literary and Linguistic Computing* 20 (2005): 189–204.

44 Windram, "Dante's *Monarchia*," 460.

45 See Robert O'Hara and Peter Robinson, "Computer-Assisted Methods of Stemmatic Analysis," in *The Canterbury Tales Project Occasional Papers*, vol. 1, ed. Norman Blake and Peter Robinson (Oxford: Office for Humanities Communication, 1993), 61.

first work with a computer-generated, unrooted tree before determining the direction of descent and the ancestral root.[46]

As promising as phylogenetics seems for the field of textual criticism, it has not been embraced wholeheartedly by textual critics. Some hesitation to rely on these programs stems from the fact that textual transmission presents challenges, such as contamination or coincidental agreement (that is not due to genealogy), that seem too complex for a computer program to solve.[47] But behind the objections over technical difficulties lies the suspicion that the use of stemmatics represents a move to supplant traditional philological methods and, more significantly, philologists.[48] To allay this suspicion, advocates of the New Phylogeny have emphasized that cladistics supplements, not replaces, traditional methods.[49]

46 Ibid., 61–2.

47 See Ralph Hanna, "The Application of Thought to Textual Criticism in All Modes—with Apologies to A. E. Housman," *Studies in Bibliography* 53 (2000): 165–7, on the failure of a program to detect coincidental agreement in the scribal mishandling of the word "wright," thus inferring a genetic relationship between texts when they may simply agree by coincidental convergence: "No branching tree model of transmission is prepared to deal with 'convergence,' for it is not the vertical procedure branching trees outline. Finding it potentially prevalent anywhere in a textual tradition renders that tradition not susceptible to genetic treatment." (167)

48 Hanna, "Application of Thought," 165: "*The Canterbury Tales Project* editors have done, by their lights, everything right and, as a consequence, have it all wrong. There is no family-tree principle, equitably and rationally applied, which can construe the reading of the isolated Ld^2 Ln Ry^2 as the reading of O, the copy provided the archetypal scribe(s) of Chaucer's 'Wife of Bath's Prologue.' And yet that reading is, on the basis of all evidence, most assuredly what Chaucer did write. One does not learn that from a machine or rule, but precisely on the basis of what Bentley called 'ratio' and Housman 'discriminating faculty'. . . . Moreover, an analysis of this variation should further undermine one's confidence that a 'machine' might generate anything like 'the truth' here—or anywhere else in the 'The Wife of Bath's Prologue' or *The Canterbury Tales*." See also Alex Jones, "Properties of a Stemma: relating the Manuscripts in Two Texts from *The Canterbury Tales*," *Parergon* 18 (2001): 37, regarding the tenuousness of "genetic relationships" inferred by phylogenetic methods: "[this is] a problem which arises afresh with the claims of Barbrook et al. that computer-administered clustering and dividing algorithms can do some of the work of textual scholars for them." See also Neil Cartlidge, "*The Canterbury Tales* and Cladistics," *Neuphilologische Mitteilungen* 102 (2001): 139ff., on the difficulty of coding variants for computer analysis.

49 Michael Stolz, "New Philology and New Phylogeny: Aspects of a Critical Electronic Edition of Wolfram's *Parzival*," *Literary and Linguistic Computing* 18 (2003), 142, "Models of this kind cannot replace traditional philological methods of analysis, but they do offer a meaningful supplement to them." Also Windram, "Dante's *Monarchia*," 444: "A realistic position is probably that these phylogenetic methods offer a valid approach to manuscript analysis, but that their real value comes in their power to assist the textual scholar rather than to attempt to replace the scholar's expertise."

A Phylogeny of the New Testament?

In many ways, the field of New Testament textual criticism seems perfectly poised to adopt cladistics to map the phylogeny of the New Testament. First of all, the thousands of extant sources for the New Testament and their multitude of variant readings present a body of data that makes the use of computer processing and analysis practically an absolute necessity.[50] Second, although Bengel stated that, ideally, textual critics would create a table of genealogical relationships between the texts,[51] and Westcott and Hort emphasized that "all trustworthy restoration of corrupted texts is founded on the study of their history, that is, of the relations of descent or affinity,"[52] a larger, more comprehensive genealogy of New Testament texts has never been realized or even thought possible.[53] Lachmannian stemmatics simply could not account for the amount of data and complexity of the NT textual tradition.[54] Could phylogenetics provide exactly the method New Testament textual critics have been looking for? Would not a tree of New Testament texts, like the Tree of Life, present the ultimate answer to the questions of descent, genealogy, and origins posed by scholars for the past three hundred years?

Yes, and no, according to New Testament textual critics. If phylogenetics represents the digital, and traditional philology and Lachmannian stemmatics the analog age of textual criticism, then the current state of New Testament textual criticism is a hybrid of both. This hybrid method is embodied by the Coherence-Based Genealogical Method (CBGM), which makes use of both computer analysis and traditional text-critical methods.[55] CBGM was created by Gerd Mink

---

50 See David C. Parker, "The Text of the New Testament and Computers: the International Greek New Testament Project," *Literary and Linguistics Computing* 15 (2000): 32–3, "The huge amount of material to be studied, the complexity of working in a number of ancient languages at once, the complicated textual situation, and the hardly less complicated history of research, all conspire to make the New Testament appear too beguilingly the supreme challenge for the use of computers in textual work."

51 Johann Albrecht Bengel, *'Η ΚΑΙΝΗ ΔΙΑΘΗΚΗ Novum Testamentaum Graecum* (Tübingen: Cotta, 1734), "Introductio," § 29; see Chapter 1.

52 Brooke Foss Westcott and Fenton Anthony Hort, "Introduction" in *The New Testament in the Original Greek*, vol. 2 (Cambridge: Macmillan, 1882), 40; see also Chapter 2.

53 See Ernest C. Colwell, "Genealogical Method: Its Achievements and Its Limitations," *Journal of Biblical Literature* 66 (1947): 109–33; see Chapter 2.

54 See Chapter 2.

at the Münster Institute for New Testament Textual Research (Institut für Neutestamentliche Textforschung—INTF) and is the method now used to create the *Editio Critica Maior* (*ECM*) and revisions to the Nestle-Aland *Novum Testamentum Graece* (NA). It is a method so impressive in its promise and impact that David Parker has stated: "I venture to predict that the Coherence-Based Genealogical Method and the use of phylogenetic software will be equal to any past advance in their significance."[56]

Like phylogenetic methods, CBGM also creates stemmata using computer technology, but its stemmata are created by subjective, philological judgments based on the traditional canons of textual criticism. Because of this, it may be easier to think of CBGM not as a method, but rather as an application that textual critics can utilize to generate results based on whatever philological method they choose.[57] As with PAUP and SplitsTree, the stage is set by inputting full transcriptions of texts (for example, of all extant, continuous Greek manuscripts[58] of the book of James) in a computer program. Once this has been done, three types of coherence can be evaluated: pre-genealogical, genealogical, and stemmatic coherence, in that order.[59]

Pre-genealogical coherence is the degree of agreement between two texts. For example, if text *a* and text *b* agree 90 out of 100 times at points of variation, then the pre-genealogical coherence between them is 90 percent.[60] The greater the pre-genealogical coherence, the more likely the

55 For a thorough introduction to the method, see Gerd Mink, "Eine umfassende Genealogie der neutestamentichen Überlieferung," *New Testament Studies* 39 (1993): 481–99; Mink, "Problems," and the presentation available at the INTF website: "The Coherence-Based Genealogical Method: Introductory Presentation" (Münster: University of Münster Institute for New Testament Textual Research, 2009), accessed May 14, 2013, http://egora.uni-muenster.de/intf/service/downloads_en.shtml.

56 David C. Parker, *An Introduction to the New Testament Manuscripts and Their Texts* (Cambridge: Cambridge University Press, 2008), 178.

57 David C. Parker, *Textual Scholarship and the Making of the New Testament* (Oxford: Oxford University Press, 2012), 98–9.

58 CBGM has so far been used primarily with continuous Greek manuscripts, to the exclusion of fragments, lectionaries, patristic citations, and so on. See Mink, "Problems," 19 (note 18), 78 and id., "Guide to Genealogical Queries," website of the University of Münster Institute for New Testament Textual Research, section 5, accessed May 15, 2013, http://intf.uni-muenster.de/cbgm/guide_en.html: "Local stemmata are only included for variants assigned to the first hands of Greek continuous text manuscripts."

59 See especially Mink, "Problems," 32–3.

texts are closely related. However, pre-genealogical coherence does not establish the direction of the relationship, that is, which text more probably gave rise to the other. It is only when a textual critic looks at the 10 points at which *a* and *b* do not agree that the quality of their relationship can be determined. At each point of variation, the textual critic determines whether it is more likely that *a* gave rise to *b*, *b* to *a*, or if the relationship at that point is unclear. If 8 out of the 10 points of variation attest to *a* giving rise to *b*, then it is most likely that the genealogical relationship between the two is $a \rightarrow b$. At each point of variation, then, a local tree or stemma of texts can be created based on pre-genealogical coherence and the direction of relationships ("textual flow") assessed between all available texts. The greatest degree of *genealogical* coherence is achieved when the optimal local tree or substemma is created to account for pre-genealogical coherence and textual flow.

Stemmatic coherence is greatest when all local trees are accounted for and a global stemma of the entire work illustrates the general textual flow based on both pre-genealogical and genealogical coherence, while at the same time illustrating the simplest network possible, that is, following the rule of parsimony.[61] Throughout the process of creating local and global stemmata, the textual critic practices what is called the "iterative process," that is, checking each stage of the method against the other to ensure that the results are self-consistent and intelligible.

Because CBGM considers both the relatedness of texts through agreement (pre-genealogical coherence) and textual flow at *every* point of variation, using full transcriptions, it combats two phenomena that were fatal to Lachmannian stemmatics: contamination and coincidental agreement. The overall coherence of agreement combined with the consistency of textual flow helps to indicate points of contamination. For example, when *a* and *b*, with a high degree of pre-genealogical coherence, show that the textual flow is predominately $a \rightarrow b$, the points at which readings show the flow in the opposite direction indicate possible

60 The total number of points of variation comparable between two texts varies depending on which pair is being evaluated. Mink, "Problems," 32: "The relevant factor in the assessment of the quality of the coherency are the percentage values of the agreements between each of these witnesses and its closest relatives, and not the absolute values, as the number of comparable places of variation changes from pair to pair of the witnesses to be compared."

61 Gerd Mink, "Editing and Genealogical Studies: the New Testament," *Literary and Linguistic Computing* 15 (2000): 55: "a global stemma must be the simplest explanation of all the relations found in the local stemmata." Local stemmata are also formed on the rule of parsimony.

contamination. Since CBGM allows for multiple ancestors, contamination can be accounted for in the global stemma. And, even with contamination, the overall genealogical coherence can be very high and reflect the most likely history of transmission, regardless of contamination.[62]

Coincidental agreement also becomes detectable because of pre-genealogical and genealogical coherence. If *d* contains a reading identical to one in *a*, and *a* has been shown through agreement and textual flow to be an ancestor of *d*, then the identical readings are a result of genealogy and not coincidence. If, however, *d* is not shown to be a descendant of *a* in local stemmata (and, thus, the global stemma), then the reading in *d* is identical to *a*'s coincidentally.[63]

It should be obvious by this point that CBGM is not possible without a computer database and software, but at no point does a computer program make text-critical decisions. The scholars at Münster are adamant that philologists be at the helm throughout the entire process. As Klaus Wachtel, a prominent scholar at INTF, stated, "A computer is not a New Testament philologist."[64] So, also, Mink explains his goals in creating the method:

> One of my purposes was to avoid statistical methods that do not allow a continuous and complete control of the relation of the basic material to the final results during the entire process following a consistent model of how the text has been transmitted. If the route from the basic material to the result is not controlled and completely understood, it is not possible to know the restrictions limiting the interpretation of the results.[65]

In the same paragraph, Mink states another major difference between CBGM and phylogenetic methods:

> Some procedures, for instance, cladistic analysis, provide diagrams similar to a tree, but the branches have no direction, so it is not possible to interpret them in terms of genealogy. It is valid as a

---

62 See Mink, "Problems," 49ff.

63 Ibid., 38ff.

64 At an annual meeting of the Studiorum Novi Testamenti Societas, quoted in Parker, *Textual Scholarship*, 96.

65 Mink, "Editing," 53.

> principle that if no genealogical data have been put in, there is *no* genealogical output.[66]

In a study comparing CBGM and cladistic analysis by using them on the same group of manuscripts (the Harclensis Group) and evaluating their distinct results, scholars from INTF and The Canterbury Tales Project also highlighted several important differences between CBGM and phylogenetic methods.[67] Beyond the fact that CBGM relies on philological decisions regarding the genealogy of readings and includes these decisions in a rooted (directional) tree, these differences include: (1) CBGM creates stemmata of *readings*, not manuscripts, unlike with current usage of cladistics; and (2) CBGM dispenses with hypothetical ancestors (hyparchetypes) that stand in as lost ancestors, whereas cladistics does not.[68]

These differences seem to set CBGM at quite a distance from phylogenetic stemmatics. What is most interesting, however, are not the technical methodological differences but the different goals reflected by them. The scholars at Münster, who include editors of the *ECM* as well as the NA, are not interested in unrooted trees because their primary goal is reaching the *Ausgangstext*.[69] Ancestry and the ancestor, however it is defined, are still the major preoccupation.[70] Furthermore, the philological

---

66 Ibid. Emphasis original.

67 Matthew Spencer, Klaus Wachtel, and Christopher J. Howe, "The Greek Vorlage of the Syra Harclensis: A Comparative Study on Method in Exploring Textual Genealogy," *TC: A Journal of Biblical Textual Criticism* 7 (2002), accessed May 16, 2013, http://rosetta.reltech.org/TC/v07/SWH2002/. See also Klaus Wachtel, "The Coherence-Based Genealogical Method: A New Way to Reconstruct the Text of the Greek New Testament," in *Editing the Bible: Assessing the Task Past and Present*, ed. John S. Kloppenborg and Judith H. Newman, Resources for Biblical Literature 69, (Atlanta: Society of Biblical Literature, 2012), 123–9.

68 Spencer et al., "The Greek Vorlage," sections 39–43.

69 The editors of the *Editio Critica Maior* (*ECM*) do not claim that the text they present as the baseline text is *the* initial text, however, only a hypothetical *Ausgangstext* A. See Barbara Aland et al., eds., *1–2 Peter, Novum Testamentum Graecum: Editio Critica Maior* (Stuttgart: Deutsche Bibelgesellschaft, 2000), 23n4: "The constructed text of an edition represents the hypothetical reconstruction of this initial text." Quoted in Michael Holmes, "What Text Is Being Edited? The Editing of the New Testament," in *Editing the Bible*, ed. Kloppenborg and Newman, 103: "This means that the initial text is the *goal* toward which the editors strive, and the lead text of the *ECM* is the *closest approximation* to it that can be achieved in the present circumstances and conditions." Emphasis original.

70 It must be stated, however, that in the *Editio Critica Maior* and the online version of the twenty-eighth edition of the Nestle-Aland *Novum Testamentum Graece*, the history of transmission plays a far larger role visually, on the page and screen, than it has ever before. This gives readers much more freedom to peruse the variants and make their own decisions regarding the baseline text.

criteria used to reach the initial text remain the same canons espoused by Bengel, Griesbach, and others: shorter readings are preferred, *lectio difficilior potior,* and so on.[71]

What is more, Wachtel, during an interview, explained INTF's rejection of phylogenetic methods by first stating that the transmission of the New Testament cannot be described as evolution:

> And there are some fundamental differences: one is, we are looking at a kind of, yes, one might call it a degeneration and not an evolution. Of course, the persons who think that the Majority Text is the original one, they might think that it was an evolution. But, no, not even they because they say that what we have in the Majority Text was already there in the beginning.[72]

This reflects a traditional view of the history of transmission, rather than the view of narrative textual criticism, which celebrates variants as a window to a diverse Christian history.[73]

All in all, despite technological advances and access to comprehensive sets of data, CBGM seems to reflect a very conservative and traditional approach to manuscripts and texts. So Parker reflects that CBGM may actually be an application rather than a method:

> My main question is one of its name. I find myself wondering whether it is a new method, or the application of a new tool to make an old method—namely Lachmannian stemmatology—work properly. Perhaps it should be called the Coherence-Based Genealogical Application. Setting the technology aside, on what is its claim to be a new method based?[74]

---

71 This is not to imply that these canons are followed slavishly. See Holger Strutwolf, "Scribal Practices and the Transmission of Biblical Texts," in *Editing the Bible*, ed. Kloppenborg and Newman, 139–60.

72 Klaus Wachtel, in an interview with the author, including Gerd Mink, conducted in Münster, Germany, March 10, 2011. For an excerpted transcription of the interview, see Appendix 1.

73 See Chapter 3.

74 Parker, *Textual Scholarship*, 98–9.

The greatest claim CBGM has as a new method is its comparison of texts at every point of variation and inferring relationships through agreement rather than difference.[75]

It is in fact these new aspects that CBGM brings to the field that make it surprisingly *like* phylogenetics, despite insistence from Münster that it is unrelated to evolutionary biology and despite a consistent faithfulness to its traditional, German roots in the eighteenth and nineteenth centuries. Phylogenetics represents a major shift in systematics or the science of classification. Rather than define and group species together through a priori or inductive categories of characteristics (Aristotelian and agglomerative classification), through ranked hierarchies of groups reflective of a divine or natural order (Linnaean), or through shared adaptations (eclectic systematics) or overall similarity (phenetics), phylogenetics classifies species *only* based on evolutionary history and relatedness. It fully reflects a post-Darwinian understanding of speciation, although it took almost a century after Darwin to reach a clear, working methodology.[76]

Up until Darwin, classification was preoccupied with the *dissimilarity* of organisms, evincing the essentialism behind scientific understandings of species. An individual finch, for example, was only a representative of the species, which was defined by all the characteristics assigned to the abstract concept of *finch*. The species finch was *essentially* different from the species wren. Darwin, however, was interested in the individual finch and how it differed from its fellow finches. For Darwin, there was no essential finch, and the abandonment of essentialism allowed him to see the *similarity* of all organisms.[77] This led him to the realization that all organisms on Earth must be related, changing and branching away from each other through the passage of time.

Cladistics embraces and emphasizes the relatedness of all organisms through its building of trees and the Tree of Life. It also insists on an

75 Even so, ibid., 99: "I feel unable to decide on the matter, but on the results I am clear. The presentation of the textual flow showing the relationship between the many states of the text leads us to view the business of New Testament philology in a new light."

76 For a brief history of taxonomy and classification, see Lecointre and Le Guyader, *Tree of Life*, 11–8. See also the discussion of eighteenth- and nineteenth-century methods of classification in Chapter 1.

77 See Lecointre and Le Guyader, *Tree of Life*, 13: "Darwin's nominalism enabled him to discover the hidden value of the variation that had previously been considered negligible. From this moment on, the question was no longer 'Why are living organisms different from each other?' but rather 'Why, despite all this variation, are organisms still similar?'"

evolutionary understanding of speciation by creating trees based only on true homologies, that is, shared characteristics due to shared ancestry, and not homoplasies, shared characteristics due to convergent evolution. Because of this, phylogenetic classification often contradicts traditional groupings of animals based on superficial morphological similarities, for example, into the categories of "mammal," "reptile," and "bird," as discussed earlier. Furthermore, due to advances in genomics, it is now possible to analyze entire DNA sequences of different organisms in order to compare comprehensive sets of data, although molecular phylogenetics faces some of the same limitations as morphological phylogenetics.[78]

The similarities with CBGM are striking. Bengel's proposition that readings be sorted into "families, tribes, and nations" set the agenda for textual classification for almost three centuries. Like the Linnaean system, it hypothesized that readings could be organized into ranked groups based on distinguishing features.[79] As with Linnaean classification, this methodology is not necessarily flawed per se.[80] But the difficulties due to contamination and an enormous amount of data drove textual critics to understand readings in a way comparable to biological essentialism. The most tenacious grouping of texts, still used in textual criticism exercises today in seminaries and universities, just as the categories of "mammals" and "reptiles" are still taught in grade school, belongs to Westcott and Hort. Their descriptions of not only the essential philological markers but even personalities of the Western, Alexandrian, and Syrian text types have left an indelible mark on text-critical imaginings of the history of New Testament textual transmission.[81]

---

78 Hervé Phillipe, Henner Brinkmann, Dennis V. Lavrov, et al., "Resolving Difficult Phylogenetic Questions: Why More Sequences Are Not Enough," in *Public Library of Science (PLOS): Biology* 9 (2011): http://www.ncbi.nlm.nih.gov/pmc/articles/PMC3057953/, accessed May 21, 2013.

79 See Chapter 1.

80 The Linnaean system's arrangement of hierarchical groups is based on an essentialist understanding of species and on a priori exigencies, e.g., the need to have seven ranked categories, because seven is a divine number. See Marc Ereshefsky, *The Poverty of the Linnaean Hierarchy: A Philosophical Study of Biological Taxonomy* (Cambridge: Cambridge University Press, 2001) 47 and Lecointre and Le Guyader, *Tree of Life*, 12.

81 Parker, *Textual Scholarship*, 83: "What I will concede to be compelling about Westcott and Hort's view is that they describe the textual process so very vividly. Such a strong and simple narrative, with certain indisputable facts, notably the dominance of a single form of text by the time we get to the later Byzantine period, is bound to be attractive. . . . It is confident and convincing. It has been accepted by most people for a long time. And yet we are left—or should be left—with the feeling that the theory does not deserve the reverence which has been accorded it." See Chapter 1.

Even recent developments in textual classification rely upon essential differences and singular readings. This is evident in both the Claremont Profile Method and the Comprehensive Profile Method, which, as their names attest, base their classification on distinctive readings (a profile), witnesses as representatives of a group, and defining group characteristics.[82]

CBGM represents a major shift away from the use of traditional text types, which had already begun at Münster through the work of Kurt Aland and Barbara Aland.[83] Rather than begin with the differences between texts, the first step in establishing genealogical relationships with CBGM emphasizes the *similarities* between all texts:

> All surviving witnesses are related to each other and there is *coherence* within the entire tradition. For all the witnesses closest relatives can be found. Between a witness and its closest relatives there is the highest degree of coherence. These closest relatives in turn have their own closest relatives, so that chains of *coherencies* develop.[84]

The relatedness of all witnesses due to the history of their transmission (or evolution) is the first assumption of CBGM. A textual critic might separate texts into groups and families after building local trees and the global stemma, but these categories are not an intrinsic part of the method, and texts are not considered essentially one text type or another. Because of this, CBGM is apt to turn up some surprising results that contradict traditional text type groupings, just as phylogenetics has in taxonomy.[85]

---

82 Bart D. Ehrman, "The Use of Group Profiles for the Classification of New Testament Documentary Evidence," in *Studies in New Testament Textual Criticism*, New Testament Tools and Studies 33 (Leiden: Brill, 2006), 33–56.

83 Kurt Aland and Barbara Aland, *The Text of the New Testament: An Introduction to the Critical Editions and to the Theory and Practice of Modern Textual Criticism*, 2nd ed., trans. Erroll F. Rhodes (Grand Rapids, MI: Eerdmans, 1995), 332–7.

84 Mink, "Problems," 32. Emphasis original.

85 Tommy Wasserman, "Criteria for Evaluating Readings in New Testament Textual Criticism," in *The Text of the New Testament in Contemporary Research: Essays on the Status Quaestionis*, ed. Bart D. Ehrman and Michael W. Holmes, 2nd ed., New Testament Tools, Studies and Documents 42 (Leiden: Brill, 2013), 605: "Some of our preconceived notions of the relationship among textual witnesses will undoubtedly be turned on their heads. . . . the CBGM has made the conception of text types problematic, at least as applied to the Catholic Letters. The traditional recensional hypothesis makes sharp distinctions among text types and their corresponding geographic locations. These notions seem incompatible with the new results. First, manuscripts from different text types show up in unexpected places, both in the global stemma and in textual-flow diagrams in substemmata, indicating textual interaction and ancestry beyond the traditional borders of the text types."

Like phylogenetics, based on a Darwinian understanding of the Tree of Life and the similarity of all organisms, CBGM does not create categories or sort texts into them based on singular features. Instead, because of technological advances, CBGM, like molecular phylogenetics, can compare comprehensive sets of data and build trees at every single point of variation. Like the finch as observed by Darwin, a text is no longer a representative of an abstract type but an important witness in its own right.[86]

These similarities are fundamental to both phylogenetics and CBGM, representing their shared ideology rather than simply "some superficial similarities," such as the creation of trees, the rule of parsimony, and an iterative process for checking results.[87] Although it may at first appear that "the two methods are fundamentally different in their aims and the ways they achieve these aims,"[88] they do in fact share the same view of their subjects of study, the nature of the resulting stemmata, and the limits of their research.

Both CBGM and phylogenetics make no claim regarding the historical accuracy of their stemmata. CBGM explicitly states as much in their introductory presentation:

> We cannot find out the exact ways of transmission ("how it really was"). We cannot construct a stemma of manuscripts. We do not know the manuscripts/texts/variants which are lost—the links between preserved copies. We can search for the structures within the development of *texts*.[89]

---

86 The same point is made by Wasserman, "Criteria," 606: "The CBGM gives priority to the role a witness plays in the genealogy of the text rather than its preconceived type."

87 Spencer et al., "The Greek Vorlage," section 39: "There are some superficial similarities between the Genealogical Method (GM) and maximum parsimony (MP). Both work with single passages of variation rather than a summary matrix of the distances between each pair of manuscripts, and use the principle of Ockham's razor in choosing between alternative stemmata by minimizing the number of changes of reading." On the iterative process in phylogenetics to account for convergent evolution (analogous to coincidental agreement in textual criticism), see Lecointre and Le Guyader, *Tree of Life*, 34: "A tree is phylogenetic if it enables the researcher to detect a posteriori the similarities inherited from a common ancestor (homologies) and the false similarities (homoplasies). In particular, a cladogram allows us to identify our successes (supported homologies) and our errors (homoplasies) among those characters we initially assumed homologous."

88 Spencer et al., "The Greek Vorlage," section 40.

89 Mink, "Coherence-Based Genealogical Method," slides 35–6, accessed May 24, 2013, http://www.uni-muenster.de/INTF/cbgm_presentation/download.html.

Phylogenetics also eschews any claim to illustrating an actual history of events:

> Choosing the most parsimonious tree has nothing to do with the so-called parsimony of evolution. We will never know the complete sequence of events that have led to the present state of nature. Choosing the most parsimonious tree is simply choosing the most consistent tree, the one that involves the fewest number of *ad hoc* hypotheses of multiple changes to resolve character conflicts.[90]

Both methods avoid any historical narrative to explain their results or, indeed, to have any part in the construction of their stemmata.

For the scholars at Münster, this is because of their explicit assumptions regarding scribes and scribal work:

> Some essential assumptions are considered more probable than their contrary (without excluding that the contrary might happen at some points): (i) a scribe wants to copy a manuscript with fidelity; primarily the scribe does not want to create new readings; (ii) if the scribe introduces other readings, they come from another source (normally a manuscript); (iii) if the scribe uses more than one source, few rather than many sources will be used; and (iv) the source copies have closely related texts rather than less related ones.[91]

In other words, no hypotheses concerning a scribe's theological, social, or political motivations will be made. The advocates of CBGM base their assumptions regarding scribal practice on a fundamental tenet of the method: the entire tradition is *coherent*:

> There can be no doubt about the existence of coherence between the states of text that have survived to our day. The high degree of coherence can be explained only by the serious and mostly successful efforts of the scribes to copy their exemplars as carefully as possible.[92]

90 Lecointre and Le Guyader, *Tree of Life*, 30.

91 Mink, "Editing and Genealogical Studies," 52.

92 Wachtel, "Coherence-Based Genealogical Method," 130.

Possible cases of scribal practice contrary to CBGM's assumptions are understood as exceptional,[93] but even then no conjectures are made as to *why* the scribe behaved in such a way for the simple reason that there is no way of confirming such conjectures:

> the tendency is [for scholars] to say they have enough room [with texts] in the second century to play with variant hypotheses, what may have happened in the evolution of the text. And *The Living Text of the Gospels* by David Parker or *The* [*Orthodox*] *Corruption* [*of Scripture*] by Ehrman and so on [are examples of this]. But what is the evidence? What is the evidence? Our experience is that nearly all texts have very close relatives.[94]

If there is any "story" to be told by manuscripts, the discovery of it may be a valid enterprise—but it is not a part of CBGM, nor really of textual criticism proper.[95]

True phylogenists also insist that there can be no narrative, no cause-and-effect story to explain the Tree of Life. The first reason, unique to its field (as scribal faithfulness is unique to the field of textual criticism) is based on the fact that cladists deal with stretches of time too long for the human mind to comprehend:

> We *invent* these stories, after the fact, to justify the history of life according to our own prejudices. . . . All we have are two isolated observations—the apparent absence of dinosaurs 65 million years ago, and evidence for a catastrophic phenomenon, such as the

---

93 Mink, "Coherence-Based Genealogical Method," slide 107: "As a rule, these assumptions [about scribal practice] will explain the evidence sufficiently. Where they do not, we have to look for an exceptional scenario matching an exceptional case!"

94 Gerd Mink, interview with the author in Münster, Germany, March 10, 2011.

95 Wachtel, interview with the author in Münster, Germany, March 10, 2011, on the work of David C. Parker, which spans the narrative and traditional textual critical divide: "So these manuscripts no doubt have stories to tell, and they are very interesting artifacts under many perspectives, art-historical, paleographical, cultural-historical, no doubt about that. But we are working together now on the Gospel of John, and will apply the Coherence Method, so he is, yes, open to both aspects. If he performs something like narrative textual criticism, well, then, that is one aspect of the history of the text that he likes to focus on—under a perspective that does not exclude *textual criticism*." Emphasis added.

> impact of an asteroid, at around the same time. There can be no certain link between the two. Geological time admits no narrative in which causes can be linked with effects.[96]

If textual critics such as Wachtel object to phylogenetic methods for use on texts because they find teleological or progressive evolution ill-suited to analyze an increasingly "corrupted" and "contaminated" text, then their objection is groundless. True cladists reject all such evolutionary storytelling.[97]

The second reason phylogenists reject narrative, like proponents of CBGM, is the lack of evidence and confirmation for any narrative of evolutionary relationships:

> Conventional stories about evolution, about "missing links," are not in themselves testable, because there is only one possible course of events—the one implied in the story. . . . You can either subscribe to the story, or not—there are no alternatives.[98]

In this way, "historical" paleontology, in the view of strict cladists, does not really have a place at "the high table of science" because its results are not testable:

> You cannot go back in time to watch the dinosaurs become extinct or fishes crawl from the slime to become amphibians. More pointedly, you cannot . . . go back in time to see what other things *might* have happened instead, had circumstances been slightly different. What if the asteroid missed Earth, sparing the dinosaurs? What would have happened if the fishes decided to stay underwater after all? . . . In strict, scientific terms, such questions are meaningless. The problem is that what we see before us is the result of a once-only experiment in history. Because

96 Henry Gee, *In Search of Deep Time: Beyond the Fossil Record to a New History of Life* (New York: Free Press, 1999), 2. Emphasis original.

97 See also Richard Dawkins, "The Conceit of Hindsight," in *The Ancestor's Tale: A Pilgrimage to the Dawn of Time* (New York: Houghton Mifflin, 2004), 1–11.

98 Gee, *In Search of Deep Time*, 7.

> it happened only once, it is not accessible to the reproducibility scientists usually require. This is not possible in paleontology except in our imaginations.[99]

The same critique could be made by traditional textual critics of narrative textual critics with the replacement of a few terms—"dinosaurs" with "Donatists" and "asteroid" with "orthodoxy," for example.

The strong place from which phylogenists and those using CBGM make this argument is one of testability, making use of the evidence at hand, and a consistent principle, the rule of parsimony. Again, the following statement made by a cladist could be made by a CBGM textual critic using analogous terms:

> Cladistics looks only at the pattern of the history of life, free from assumptions about the process of the unfolding history. . . . More than a set of techniques, but less than a science in its own right, cladistics is a way of 'seeing,' of looking at the products of evolution as they are, and not how we would like them to be. An important aspect of cladistics, as in all science, is testability. In cladistics, you are asked to find the most likely way in which a set of organisms is related to one another—to estimate, in other words, their relative degrees of cousinhood. This is done with a cladogram. . . . there is more than one way of drawing a cladogram that links them up. What this means is choice: if there is more than one way in which organisms can be cousins, you are forced to consider the alternatives, and you must find a way of evaluating them all. How can these alternatives be evaluated? The central test is as old as science—older: Occam's razor, or the Principle of Parsimony.[100]

Again, textual criticism uses scientific and biological discourse to understand, analyze, and interpret its subject. The CBGM presents a rather clinical perspective of the history of textual transmission, and some have doubted the wisdom of this. Eldon Epp, ever the champion of narrative textual criticism, writes, "One deficit of the electronic application, therefore, is the absence of narrative about the contextual influences that shaped the text

99 Ibid., 8.

100 Ibid., 6–7.

during its long history." But at the same time Epp also questions the "objectivity" of CBGM purported by its discourse and technical presentation.[101]

As has been discussed, scholars using CBGM are the first to point out that it is a method based on traditional philology and the instincts of a textual critic. Be that as it may, there is no place for a narrative of scribal editing or rival Christological beliefs to explain the stemmata produced. CBGM is, in essence, a scientific method: a testable, iterative process, without narratives, based on (subjective) human observation and interpretation.[102]

The Text as a "Carrier of Meaning"

Scholars using CBGM are very careful to make clear the distinction between a manuscript and a text—they are, after all, not creating stemmata of manuscripts but of texts. As Pasquali famously stated, *recentiores non deteriores,* newer manuscripts are not necessarily inferior because they may bear older texts.[103] For this reason, CBGM maps the relationship between "states of text":

> The state of text in a manuscript has to be clearly distinguished from the manuscript as artifact with its paleographical and codicological

101 Eldon J. Epp, "Textual Clusters: Their Past and Future in New Testament Textual Criticism," in *Text of the New Testament,* 552. The quotation is part of an assessment of CBGM's objectivity and subjectivity: "Computer programs are viewed often—but naively—as simple and thoroughly objective, but, on the contrary, it is clear that the CBGM involves heavy elements of human judgment and interpretation both before and after the data are submitted to computer analysis for tabulation and further refinement in the reiterative process. It would appear, however, that such extensive iteration focuses increasingly on the mechanical relationships of texts as seen in their mutual agreements, and less and less on the observable characteristics of the texts themselves. One deficit of the electronic application, therefore, is the absence of narrative about the contextual influences that shaped the text during its long history. Such considerations, of course, are inherent in various criteria for isolating the prior reading in each local stemma, and these criteria involve a considerable measure of subjectivity. Is not the result, then, a process with less objectivity than at first appears?" (551–2). In any case, Epp foresees these as issues to be dealt with in the future of CBGM and the field in general.

102 CBGM as a use of methodical evaluation of subjective analysis is stated in Mink, "Editing and Genealogical Studies," 56: "My intention is to reduce subjective evaluation of readings and witnesses and to strengthen the external criteria. But internal criteria are important now as before. They are not really restrained, but applied in a new methodical environment. The external criteria become more powerful and reliable if we pay close attention to the principle of coherence and to the role each witness plays in the textual flow."

103 See Giorgio Pasquali, *Storia della tradizione e critica del testo,* 2nd ed. (Florence: Le Monnier, 1962), 41–108.

> features. The relationships analyzed by the CBGM are strictly between states of text, not between manuscripts.[104]

The description above is rather brief, and it does not actually provide an explicit definition apart from differentiating "states of text" from manuscripts, but it is revealing. That a text has a "state" and even multiple "states" may seem apparent in the field of textual criticism. After all, if a text were merely the text at hand, without different forms or states, textual criticism would be unnecessary. But if a text is not merely itself but a state of something—or if it can have several states (or even hundreds and thousands)—then what exactly is a text?

From one implicit description, a text not only has different states but also physical and non-physical aspects: scholars who are less than careful in distinguishing manuscripts from texts in their methodology will "not be consistent in discriminating the text from its *physical appearance*."[105] It seems that the text is understood in Platonic terms: behind the physical manifestation of markings on a manuscript, there lies an essence that is truly the text.

Mink's explicit definition of the text confirms this:

> I should also like to point out in this context that the notion of *text* needs a more exact definition. In textual criticism it is mostly used for what "was written down" in the manuscript, strictly speaking for the sequence of characters the copyist wrote down. On the other hand, the text should convey meaning. This aim, however, is not directly achieved by the written characters. They represent different things in different languages.[106]

Pausing here, it should be clear that Mink understands text to be a conveyer of meaning and this meaning is not completely tied to the written characters on a page. A close reading of the last sentence in the quotation also gives an oblique view of Mink's understanding of language and semiotics: language functions as a representation of things.

---

104 Wachtel, "The Coherence-Based Genealogical Method," 129.

105 Spencer et al., "The Greek Vorlage," section 14. Emphasis added.

106 Mink, "Problems," 26.

The next part of the definition goes further in differentiating the "physical appearance" of the text and the text as a conveyer of meaning:

> A copyist may grasp the sense of a text and copy this sense. Even if he intends to copy the text true to the letter from the exemplar, he may, at the level of the characters, produce variants which emerge from the meaning of the text. But a copyist may also, particularly in long or difficult sentences, copy groups of words, or line by line, while losing the overall meaning of the text. In the case of very difficult words he might even copy character by character, which may lead to readings which hardly make any sense, and even to spectacular errors.
>
> Now, the problem is that there is no obvious difference between variants introduced on purpose by copyist[s], and variants produced accidentally, as long as these variants are meaningful in the widest sense. The text as carrier of meaning is therefore modified by the variants—introduced intentionally or unintentionally by the copyist—, and there is no way of differentiating between intentional and unintentional variants.[107]

There appears to be the text at the level of the characters, groups of words, and line by line, and these may be termed the "physical appearance" of the text, which the scribe intends (it is assumed) to reproduce accurately. But there is also the text as carrier of meaning, which is what the copyist might grasp in his mind. It is, as Mink has stated, "an abstract thing."[108] If a variant (regardless of its intentionality) is meaningful (that is, can be understood and not simply gibberish), then that abstract text as carrier of meaning has been modified.

This may all seem like common sense. Of course if characters or groups of words are modified, then the text is modified! But the assumptions behind these descriptions are significant. First, the text itself holds meaning. Second, this meaning is somehow behind or hovering above the text (its carrier) so that even if a copyist is faithful to the meaning, he may be unfaithful on the level of characters or groups of words. The

107 Ibid.

108 Mink, interview with the author in Münster, Germany, March 10, 2011: "So the manuscripts themselves [are] parchment, or whatever [they] may be. The text is an abstract thing."

meaning can be distorted if the text as carrier is modified, but this implies a true state of meaning and a true state of the text.

"True" may seem like too loaded a term to use here. After all, the scholars at Münster use CBGM only to reach the hypothetical *Ausgangstext*, the initial text and not the "true" text, not the "original" text or a new *Textus Receptus*. But stating that the text is "a carrier of meaning" implies that textual scholars are not simply working with texts but with meaning, an abstract presence beyond the letters. They do not work simply with states of text looking for an initial text but also with states of meaning in search of the initial meaning, which survives somewhere in the web of modified texts, an image projected from their collective mosaic.

This becomes quite clear when Mink describes what happens to the abstract text once non-meaningful variants (gibberish or spelling mistakes) become a part of the physical text:

> If, however, a copy contains errors, i.e. readings which clearly do not make sense, these do not necessarily modify the text as a carrier of meaning. The copyist had no intention of changing the text. The copy therefore only contains an erroneous representation of the same text at the level of the characters, which nevertheless sometimes renders it unrecognisable as a carrier of meaning. But not only errors at the level of characters are possible. After all, any linguistic element can be affected by mistakes of this sort, such as errors of concord and case, also omissions of key words for the context, which happen accidentally and against a scribe's intention (*lapsus*).
>
> Even the first exemplar of the entire textual tradition, even the very autograph, may have contained such errors. The text as the actual carrier of meaning, however, would not necessarily be affected by these errors, it would only be *represented defectively*.[109]

In other words, the meaning survives, even if disfigured on a physical level. It is an entity connected but not fixed to its physical manifestation.

In biology, a "carrier" is a person who carries a particular gene that is not necessarily expressed by the person's bodily appearance or condition. This is because genetics operates on two levels, at the level of the actual

109 Mink, "Problems," 26. Emphasis added.

genetic code, the genotype, and at the level of the physical expression of that code, the phenotype. What Mink and others at Münster describe as the different aspects of the text is strikingly analogous to genes. A text, with all its aspects, is like an individual human being who carries a genetic code (the meaning) but who may represent this code in a modified way phenotypically (the characters on a page). Human beings can, of course, become disfigured or ill, but this does not affect the genetic code with which they were born.[110] The code and the meaning remain the same. The analogy becomes rather complex and contradictory when taking into account all other text-critical analogies regarding sexual reproduction and generation discussed already, but it is sufficient to state here that the concept of text in CBGM is a genetic one: it has both a genotype and a phenotype.

This is exactly the analogy used in a popular handbook on evolution, and it provides a provocative entrée into the issues at play in conceptualizing the text as a "carrier of meaning":

> For the non-biologist, the easiest way to envisage how genetics works at the molecular level is to draw on an analogy between a genome and a handwritten manuscript carefully copied by generation after generation of monks. Just as a manuscript is written in the letters of an alphabet, the genome is built from four DNA characters, A, C, T and G. In the manuscript, strings of letters make up larger units (words, paragraphs, chapters) that, when read through the human eye and brain, produce meaning. Similarly, sequences of bases in DNA make up genes that, when read through a cell's molecular machinery, elicit functions, usually by providing recipes for proteins. Changes arise in the manuscript and the genome during the process of copying. We see similar changes in both contexts: one character can change into another; one or more character[s] can be inserted or deleted; strings of characters can be moved from one part of a manuscript/genome to another or to an entirely different document/cell.[111]

110 Bracketed out from this discussion and analogy are the findings of the growing field of epigenetics, the study of heritable characteristics that are *not* coded in DNA, although further investigation of epigenetic phenomena may be useful considerations for textual critics.

111 Mark Pallen, *The Rough Guide to Evolution* (New York: Rough Guides, 2009), 136–7.

The genetic processes described above are DNA transcription, translation, and replication—for once, scientists have used the language of textual critics to describe biology! Essential to all three of these processes is the chemical attraction among the nucleotides A, C, T, and G and their complementary nucleotides (A and T are complementary, as are C and G; in RNA T is substituted by the nucleotide uracil (U)). Very briefly, DNA replication occurs when the double helix structure of DNA separates (like a zipper unzipping) and a molecule called DNA polymerase lays down complementary nucleotides on each of the separated strands. When the new strands are complete, they break away and together form a new double-helix DNA.

Both DNA transcription and translation are part of the process of gene expression. Transcription also begins with the separation of the double helix. A single strand of that unzipped DNA is "transcribed" by RNA polymerase so that nucleotides complementary to the exposed string of A, C, T, and G nucleotides form a new strand called messenger RNA (mRNA) (see Figure 4.1). Translation occurs when the newly formed mRNA breaks away, anchors to a ribosome (an organelle in a cell), and transfer RNA (tRNA), carrying an amino acid, attach to the strand at a complementary location, side by side. The amino acids carried by the tRNA are then linked in sequence and form an amino acid chain (see Figure 4.1). Once the whole sequence of RNA has been "translated" by tRNA, the amino acid chain breaks away and forms a protein, the end product.

The comparison of DNA replication with scribal copying is already implicit in the use of phylogenetics for textual criticism, where the sequence of letters in a text is coded *as* DNA for software analysis. A closer look at the analogy reveals a few things. If a scribe functions like DNA polymerase, then he responds automatically and predictably. DNA polymerase, as far as we know, does not lay down complementary nucleotides *intentionally* but through chemical reactions in order to form molecular bonds. The "mistakes" that it may make are mostly structural, where bonds may "wobble" because of the flexible nature of the DNA strand, for instance.[112] Mistakes are fixed when DNA polymerase "proofread" (another literary metaphor used in biology) the strand and react to improperly bonded nucleotides.[113] If scribes are comparable, then they are

112 Leslie A. Pray, "DNA Replication and Causes of Mutation," *Scitable (2008)*, accessed May 30, 2013, http://www.nature.com/scitable/topicpage/dna-replication-and-causes-of-mutation-409.

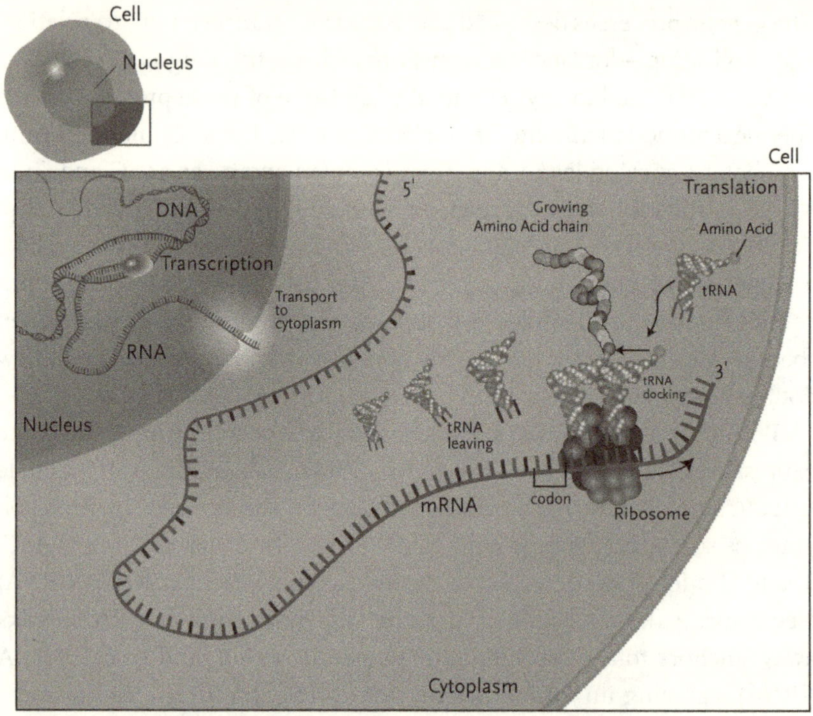

FIGURE 4.1. DNA Transcription and Translation. Image by Darryl Leja, courtesy of the National Human Genome Research Institute, www.genome.gov.

not creative agents of any sort but simply reacting to their exemplars and reproducing them automatically. Their correcting of previous mistakes is likewise reactionary and automatic.

It might seem that this type of detailed comparison takes the analogy too far, except that CBGM assumes that scribes function in exactly this way. They are termed specifically as "copyists" and not "editors."[114] It is assumed that they had no editorial agency and certainly did not take creative license in performing their tasks. As discussed above, the general "coherence" of the New Testament textual tradition serves as proof of their accuracy. The existence of multiple variants is simply a by-product of hand-copying and not evidence of any intention. After all, DNA polymerase can be as accurate as making an error only once while copying

113 Ibid.

114 See Wachtel, interview with the author in Münster, Germany, March 10, 2011.

one billion nucleotides—but even then mutations can accumulate quickly through cell division.[115]

This conception of scribes as automata and variation as unintentional, by-rote events stands in sharp contrast to the dynamic, vibrant text of narrative textual criticism, discussed in the last chapter. Indeed, the proponents of CBGM at Münster and the major advocates of narrative textual criticism represent quite different understandings of the proper reach and ken of textual criticism. The irony, however, is that both viewpoints work with the metaphor of a *living* text. Narrative textual critics perceive the text as an evolving, adapting creature—evolution on a macro level. Phylogenetic textual critics and CBGM (although the latter only implicitly) perceive the text as a sequence changing through hundreds and thousands of rounds of replication—evolution on a micro level.

In both understandings of the text, the reader—either scribe or interpreter—is less an active agent than a reactor. In the DNA-text analogy above, "strings of letters make up larger units (words, paragraphs, chapters) that, when read through the human eye and brain, produce meaning."[116] In this description, the human eye and brain are no more than a processing mechanism; they are entirely passive. Instead, all activity belong to the letters: they are read and they *produce meaning*. The human mind, like a molecule of tRNA, simply reacts to the text (mRNA), which is what causes the mind to string together the correct interpretation, the chain of amino acids.

Again, it may seem ridiculous to follow the analogy to this extreme, but the notion of the human mind operating like a molecule of tRNA is not far off the mark when considering the text as a "carrier of meaning." If the text itself contains meaning, then it is the agent of interpretation, eliciting a reaction from the human mind. The reader, like tRNA, should grasp that meaning, but there is the possibility of error in understanding, just as there is in DNA translation.

This is a very narrow conception of the human act of reading and interpretation. The text as "carrier of meaning" implies an interpretation that is automatic and predictable—or should be. This robs the human mind of agency and creativity, and it implies a normative, correct interpretation. The objection could be made that Mink and others are not

115 Pray, "DNA Replication."

116 Pallen, *The Rough Guide to Evolution*, 136.

speaking of interpretation but of scribal behavior and transcription. However, that means these scholars would say that the text is ontologically one thing for a scribe and another for a reader. This is certainly not the case because Mink states that even for the *author*, the traditional well-spring of meaning and intention, the sequence of letters carries a meaning beyond its physical manifestation: "even the very autograph may have contained such errors. The text as the actual carrier of meaning, however, would not necessarily be affected by these errors, it would only be represented defectively."[117]

At this point in the discussion, some may claim that what Mink or other textual critics state about the nature of text is inconsequential for interpreters because their primary work is not hermeneutics, nor do they elaborate a theory of language for the purpose of exegesis. They do not presume to dictate what an interpretation should be; they only offer others the tools by which interpretation can be conducted. This has been, after all, the goal of textual criticism all along, to provide a stable text that an exegete can use. It is from this understanding of the task of textual criticism that the other name of the field is derived: lower criticism. Lower criticism is supposed to provide the foundation upon which higher criticism—interpretation, exegesis, hermeneutics—is built.

However, it should become apparent upon further reflection that the tools provided for interpretation presume the type of interpretation that should be produced. When it is stated that scribal errors disfigure or modify the meaning carried by a text, the meaning, as a static, recoverable entity, becomes the assumed desideratum of the interpreter. When the task of textual criticism is not only to answer a historical question, for example, what text best explains the history of transmission, but also to claim a definitive text not just as a baseline text but as a source of authority, then textual critics' beliefs regarding language and meaning are as significant as the exegesis of interpreters using their editions. As Parker states:

> the way we think about texts does not concern just the way we think about texts. Around our view of texts revolve notions of authority, and with that such matters as feelings of guilt and freedom, power and sense of powerlessness.[118]

---

117 Mink, "Problems," 26.

118 David C. Parker, "Manuscripts of the Gospels in the Electronic Age," *Restoration Quarterly* 42 (2000): 229.

The statement of purpose given by the INTF is quite careful to maintain that the work of the institute is historical investigation:

> Our central task is to research the textual history of the New Testament and to reconstruct its Greek initial text on the basis of the entire manuscript tradition, the early translations and patristic citations.[119]

It makes no claims as to the authority of the text it reconstructs. The prefaces to the NA twenty-seventh edition and to the *ECM* also eschew any notion of reproducing the "original" text or any implicitly authoritative text, and this reflects the basic shift in textual criticism discussed in the preceding chapter.[120]

Eldon Epp has, in a similar vein, stated that "the text critical discipline per se carries with it no normative implications and imposes no theological overlay."[121] And the conversation among textual critics regarding the sometimes problematic notion of a "text" shows that certain broadly accepted assumptions about text and the authority of editions are increasingly being questioned. The conversation involving both textual criticism and literary theory began in earnest with Jerome McGann in the 1980s, when he questioned the concepts of "final intention" and "author's intention."[122] Even the visual arrangement of the text and apparatus in forthcoming digital editions of the New Testament is beginning to reflect the multiple identities of texts and the instability of the baseline text.[123]

---

119 Homepage of the University of Münster Institute for New Testament Textual Research, accessed June 3, 2013, http://egora.uni-muenster.de/intf/index_en.shtml.

120 See Holmes, "What Text Is Being Edited?" especially 94–105.

121 Epp, "Multivalence," 279.

122 Jerome J. McGann, *A Critique of Modern Textual Criticism* (Charlottesville, VA: University Press of Virginia, 1983). See also Philip Cohen, ed., *Devils and Angels: Editing and Literary Theory* (Charlottesville, VA: University of Virginia, 1991).

123 Both the *Editio Critica Maior* and the Nestle-Aland *Novum Testamentum Graece* will have digital versions that allow the reader greater access to the data used to form the baseline texts of those editions. See "The *Editio Critica Maior*," website of the University of Münster Institute for New Testament Textual Research, accessed March 28, 2015, http://www.uni-muenster.de/INTF/ECM.html; and "Digital Editions," website of the Nestle-Aland *Novum Testamentum Graece*, Deutsche Bibelgesellschaft, accessed March 28, 2015, http://www.nestle-aland.com/en/extra-navigation/digital-editions/. Even more capabilities of the digital *Novum Testamentum Graece* were demonstrated by Zeth Green and David C. Parker, "The Digital Nestle-Aland," presentation given in the session "Critical Editions of the German Bible Society: Nestle-Aland 28th Edition," Society of Biblical Literature Annual Meeting, Chicago, November 19, 2012.

Nevertheless, the notion of an underlying, recoverable meaning within the text abides in New Testament textual criticism and, with it, a narrow and normative conception of interpretation. This is true of both traditional and narrative textual criticism. For example, Bart Ehrman states:

> The importance of establishing a hypothetically "original" text has always been fairly self-evident to historians: you can't know what an author meant if you don't know what he or she said.[124]

In justifying the field, Ehrman seems to take for granted that (1) authorial intent is essential to interpretation and (2) it is possible to recover authorial intent. As a proponent of narrative textual criticism, Ehrman emphasizes the significance of variant readings and evidence of a diverse Christian exegesis through the centuries.[125] And yet, uncovering what "the author meant" by uncovering his or her words is a necessity for interpretation:

> Establishing what an author wrote is an *indispensable* first step to determining what he or she meant. Within the pages of the New Testament there are textual variations that have not yet been satisfactorily resolved and that have profound effects, not just on a word here or there, but on the entire meaning of entire books and their portrayals of Jesus. . . . These textual problems cannot simply be swept under the table and ignored. Commentators, interpreters, preachers, and general readers of the Bible *must* recognize their existence and realize the stakes involved in solving them.[126]

What exactly are the "stakes involved"? A correct and true interpretation? Do all interpreters have to base their interpretation on either a "resolved"

---

124 Bart D. Ehrman, "Text and Interpretation: The Exegetical Significance of the 'Original' Text," in *Studies in the Textual Criticism of the New Testament*, New Testament Tools and Studies 33 (Leiden: Brill, 2006), 309.

125 In fact, the above quotation is taken from a context in which Ehrman emphasizes the importance of knowing authorial intent in order to see how or why later scribes changed it. Immediately after the above quotation, Ehrman writes, "The importance of variant readings, however, has rarely been as self-evident to historians, although it is now becoming the most exciting area of study in this field. For once it is known what an author wrote, one can ask why the text came to be changed by later scribes living in different circumstances" (ibid., 309–10).

126 Ibid., 324. Emphasis added.

text or limit themselves to interpretations that take the complexities of textual criticism into account? Or perhaps Ehrman infers here what Parker explicitly states: the multiplicity of texts and readings *is* the Christian tradition; there is no one ancient, authoritative interpretation to reach because it never existed.[127] To claim that no interpretation is justified unless based upon a text that is the product of modern textual criticism would invalidate centuries of interpretation.

Yet the comparison of the text to a living organism perpetuates the notion of a singular, correct meaning of the text. In narrative textual criticism, the text is a "living, breathing" animal, with its own intention and agency, independent of the reader. In CBGM, it is a sequence of letters that carries meaning within it, like a strand of DNA that is "pure information."[128] The interpreter, in these metaphors, approaches a living entity that has its own voice and utterance, and her role is not the creation of an interpretation but to listen or react passively to the meaning within the text.

Biological metaphors have served the field of New Testament textual criticism well by providing new methodologies and intellectual paradigms with which to organize, describe, and appreciate the history of textual transmission. But, as Darwin wrote, "analogy may be a deceitful guide."[129] A living text that adapts to its environs or passes on its genetic coding faithfully for generations is a dynamic concept. But the focus on a living text neglects the only true living entity in the act of reading, transcribing, and exegesis: the human interpreter.

---

127 David C. Parker, *The Living Text of the Gospels* (Cambridge: Cambridge University Press, 1997) 203–13.

128 Richard Dawkins, *River Out of Eden: A Darwinian View of Life* (New York: Basic Books, 1995), 19: "Genes are pure information—information that can be encoded, recoded and decoded without any meaning. Pure information can be copied and, since it is digital information, the fidelity of the copying can be immense." Whether a molecule of deoxyribonucleic acid can be termed "pure information" should, perhaps, also be under discussion.

129 Darwin, *Origin of Species*, 484.

Conclusion

We're tired of trees. We should stop believing in trees, roots, and radicles. They've made us suffer too much. All of arborescent culture is founded on them, from biology to linguistics.

GILLES DELEUZE AND FÉLIX GUATTARI, *A Thousand Plateaus*

Stemmata quid faciunt?

JUVENAL, Satire 8

I sing of Marcus Niebuhr Tod
Who found the autograph of God.
Carved on a Sinaitic cliff
He found Jehovah's hieroglyph
A casual glance assured the date
As B.C. 1698 . . .

MAURICE BOWRA, "Marcus Niebuhr Tod"

WHAT IS A tree? More specifically, what is a tree in the language of phylogenetics, CBGM, and textual transmission as understood via Lachmannian genealogy? The multifarious conceptions of "tree" present a shifting constellation of elements, so that the term "tree" has been plastic enough to provide a chief metaphor for both biology and, thus, textual criticism, for the last three hundred years. The overriding utility of the metaphor lies in the connections among branches, trunk, and roots, which convey the relatedness—however understood—between texts and species. All other elements and conceptions of "tree" can be manipulated to illustrate hierarchy, direction of textual flow, degree of relatedness, narrative or even anarrativity in the case of "unrooted trees" in phylogenetics.

Graphically, textual trees or stemma present the multiple identities of what is commonly referred to as *the* New Testament, as if "it" were a unified, monolithic entity. The pointillistic image of all a tree's members vividly shows the mosaic quality of the New Testament as a historic entity and quite aptly illustrates the argument that "Scripture is tradition."[1] Thus, another strength of the tree metaphor is its emphasis on growth and change—not only because of its ramifying branches and roots but also since trees themselves grow—unlike webs, for example, which are another possible metaphor.

In the same way, the metaphor of the text as a living being has been patently useful, dynamic, and apt—so much so that at times the terminology seems no longer metaphorical. On the most basic level, a text used as a source does indeed seem to have "offspring" derived from it, sharing its characteristics and created through a pairing of the text with human agency. On a more complicated conceptual level, racial classification provides not only a method by which the vast number of texts can quickly be surveyed but also analogous elements that give textual critics a dynamic shorthand to describe texts. The style and idiosyncrasies shared by a group of texts becomes a "textual physiognomy" or "textual complexion"; texts differing slightly but significantly from these characteristics can be described as "close relatives" but not part of the same "family," and so on. And why not continue with the analogy? A text "family" has a provenance (or multiple provenances—whether they are known or not is another matter) and have seemed to textual critics to reflect the values and cultures of their environment, just as human "races" seemed to natural scientists to have colors, shapes, and mentalities reflecting the climate and environment of their native heath.[2] The close-matching analogous elements blur the distinction between the metaphor and the subject at hand.

---

1 The epigraphs come from Gilles Deleuze and Félix Guattari, *A Thousand Plateaus: Capitalism and Schizophrenia*, trans. Brian Massumi (Minneapolis, MN: University of Minnesota Press: 1987), 15; and Maurice Bowra, *New Bats in Old Belfries: Some Loose Tiles*, ed. Henry Hardy and Jennifer Holmes (Oxford: Robert Dugdale, 2005), 3. For the full poem, see Appendix 2. See David C. Parker, *The Living Text of the Gospels* (Cambridge: Cambridge University Press, 1997), 203–7. It should also be noted that not only is the New Testament not a unified entity in terms of its texts but also in the separate textual traditions of the Gospels as distinct from, for example, the Apocalypse or the Pauline corpus. These realities are represented in textual trees constructed by CBGM that create the stemma only of a particular book of the New Testament, since, of course, not all books of the New Testament are included or survive in each extant manuscript.

2 For more on the environmental theory of race not only from the eighteenth century on but also in antiquity, see Benjamin Isaac, *The Invention of Racism in Classical Antiquity* (Princeton: Princeton University Press, 2004).

In this way, it seems inevitable that textual critics should use the biological sciences in their descriptions and methodologies. It is not as though philologists in the eighteenth century, as one textual critic put it, "woke up in the morning and decided to be racist."[3] Instead, they worked and thought as part of the intelligentsia of their day. Given the broad and popular interest in the "natural sciences" and biology, then and now, and the striking parallels between textual and human genealogies and development, it seems a matter of course for textual critics to refer to textual offspring, descendants, families, adaptation, and gene pools.

And the biological models of speciation, development, and genealogy have paid off in articulating and reconstructing the textual transmission of the New Testament. Bengel's racial metaphor launched nearly three hundred years of dynamic scholarship based on the fundamental principles of categorization by group and difference. On them were built the editions of Lachmann, Tischendorf, Tregelles, Westcott and Hort, and Nestle, and from them were extrapolated Streeter's theory of local texts, Colwell's definition of text type, the Alands' categories, Epp's textual clusters, and various profiling methods.[4]

The outlier of these methods is the Coherence-Based Genealogical Method (CBGM), which has no need of text types and profiling methods because of its use of technology for quantitative analysis. What makes CBGM so intriguing—it is both notoriously difficult to describe succinctly and purported to be the greatest advance in New Testament textual criticism in recent years—is that it embraces the overarching trends in biology since the seventeenth century despite its practitioners' distancing themselves from biologists and their methods. The scholars at Münster, on the one hand, seek an ancestral *Ausgangstext*, rely on genealogical stemma, and speak in terms of "contamination" and "corruption," while on the other hand they eschew narratives of development and adaptation

3 This was said in response to a paper based on Chapters 1 and 2 of this book, "Classification, Genealogy, and Contamination: The Racialization of Texts in the Eighteenth and Nineteenth Centuries," given during a textual criticism session at the Society of Biblical Literature Annual Meeting in San Francisco, 2011. Racism is not in itself humorous, but the remark highlights the absurdity of (1) the anachronism of the term "racist" in the eighteenth century, (2) the idea that one ever *decides* to be racist or that one is either racist or not racist, and (3) the idea that racism involves a high level of self-awareness.

4 See Eldon J. Epp, "Textual Clusters: Their Past and Future in New Testament Textual Criticism," in *The Text of the New Testament in Contemporary Research: Essays on the Status Quaestionis*, ed. Bart D. Ehrman and Michael W. Holmes, 2nd ed., New Testament Tools, Studies and Documents 42 (Leiden: Brill, 2013), 519–77.

in favor of parsimonious trees reflecting only evolutionary change. In other words, CBGM uses both Buffon and Darwin, Blumenbach and Dawkins; it maintains the goals of Bengel and Lachmann while employing statistical cladistics on the cutting edge of technology.

There is no doubt that biological metaphors and methods have served New Testament textual critics well and continue to do so, as evinced by the utility and promise of CBGM and by a renewed interest in the field inspired by narrative textual criticism and the "living" text. Nevertheless, it has been a major contention of this study that ideologies of the natural sciences and biology have also curtailed and limited the scope and growth of the field.

Classification, though useful in characterizing and organizing an otherwise overwhelming abundance of texts, also oversimplifies and overlooks subtle differences. As was the case with race classifications, textual characterizations became caricatures burdened with biases so that, for example, Wescott and Hort's relied too much on their so-called "Neutral" text as the "purest" text-type. What is more, value-laden labels, including not only "Neutral" but also names such as "Oriental" and "Alexandrian," which conjure up social-cultural conceptions of intelligence, disposition, and worth, became confusing and inaccurate as the hypotheses about geographic origins that gave rise to them were obscured through time and changing scholarship. Just as "Caucasian" is still in usage as a racial category, so is "Western" as a text-type, but neither label now indicates anything related to its origins. Blumenbach named a race "Caucasian" after deciding that they originated from the most beautiful peoples of the Caucasus region. In like manner, Semler gave the name "Western" to a group of texts he determined to originate from the "Latin West." These labels now confuse matters and simultaneously propose a geographic reality as justification for the category while their wide usage obscures the fact that these justifications have been debunked.[5]

The impulse to classify the human "races" was followed by a focus on hierarchization, a search for origins, and the conception of divergence from a purported, pure origin as degenerative. By adapting a racial metaphor in the description and categorization of texts, textual critics followed suit. This is not to say that their chase after the "original text" did not also

5 The "Western" text-type has been relabeled the "D-text cluster," but the label "Western" is still in common usage in seminaries and biblical studies departments, as well as in many handbooks.

derive from a simple historical interest in the past,[6] but the description of textual variance from an originary "pure" text as degeneration and corruption betrays the underlying assumptions textual critics shared with contemporary natural scientists. The New Testament textual search for origins was arguably as fraught as the search for an ancestral human race because of its identification with holy writ. Thus, textual critics attempting to trace the genealogy of biblical texts back to an origin could imagine that their destination lay in the divine, just as the genealogy of the Gospel of Luke, which runs backwards in time, ends with, "son of Seth, son of Adam, son of God" (3:38). But unlike the Lukan genealogy, which presents the last generation of the family tree as also divine, the latest iteration of the New Testament for Lachmann, Westcott, and Hort was not an analogue to Jesus, the incarnate God, but a monstrous corruption, a bastard that paraded about under the deceptive title of *Textus Receptus*.

Thus while advancing certain theories of the history of textual transmission, the use of genealogy and racial theory metaphors of "purity," "contamination," "degeneration," and "corruption" made for a highly myopic view of the field and its possibilities. Difference signified either an obstacle to or evidence for uncovering the ancestral text and nothing more. Trees of relations were drawn up, not in order to illustrate a family history but to indicate purities and impurities. The search for an "original text" also bought wholesale the belief that such a thing could exist, that a singular, relatively unproblematic endpoint could hypothetically be reached given enough evidence and scholarly determination.

With narrative textual criticism came a renewal of life and interest in the field. The narrow goals of textual criticism, so long focused on a single point, broadened to include the diverse forms of the "living text." An evolutionary understanding of the development of the text, rather than an essentialist, classificatory conception, transformed the subject of textual criticism into material worthy of narration and story, so that variations within textual transmission presented the *mythoi* of early Christian traditions. At the same time, adopting the metaphor of text-as-living-organism and its implications brought with it the adoption of the primary *mythos* of New Testament studies, that of a primordial, free, and dynamic gospel that gradually grew domesticated and ossified. Focus on the free and living earlier texts of the New Testament entailed neglect of the Majority

6 Although it is arguable whether this interest is separable from a value-laden search for origins.

Text because of its standardized form and "tame," unchanging, even dead state.[7] Furthermore, when textual critics equate "living" with "changing," the sacred texts that are purported not to change, due to scribal practice or religious sanction, are easily assimilated to dead or, at best, zombie texts that continue to be assigned authority as scripture but do not adapt to their changing environs.

The phylogenetic model and CBGM, on the other hand, deny that any narrative can be derived from the transmission history of the New Testament. Instead, the transcription of texts for analysis transforms readings into codes, like a strand of DNA, and, in turn, scribes are understood not as theologically motivated, reactive beings but as well-trained and well-disciplined copyists, working with almost automaton-like faithfulness. The scholars using CBGM are themselves cyborg-like in combining their human, philological judgments with the power of the machine, connecting texts and ultimately generating a cyborg edition as the result of a human-machine marriage. This cyborg, borrowing from Donna Haraway's "Cyborg Manifesto,"

> [unlike Frankenstein's monster,] does not expect its father to save it through a restoration of the garden; that is, through the fabrication of a heterosexual mate, through its completion in a finished whole, a city and cosmos. The cyborg does not dream of community on the model of the organic family. . . . The cyborg would not recognize the Garden of Eden; it is not made of mud and cannot dream of returning to dust.[8]

In other words, the edition created by using CBGM is not an organic, living being with a reconstructed narrative of racial or family resemblance or communal controversies. The scholars at Münster state explicitly that

7 This has only recently begun to change. The twenty-eighth edition of the Nestle-Aland *Novum Testamentum Graece* includes more consideration of the Majority Text—henceforth represented by "Byz" for "Byzantine Text"—in the Catholic Letters since these are based on their edition in the *Editio Critica Maior*, created with the use of CBGM, which considers full collations of all continuous Greek texts, *including* the texts of the Majority Text. Before such thorough analysis was possible, there was a clear bias against Byzantine readings, as explained by Klaus Wachtel in his presentation, "The Revision of the Catholic Epistles according to the *Editio Critica Maior*," in the session "Critical Editions of the German Bible Society: Nestle-Aland 28th Edition," at the Society of Biblical Literature Annual Meeting in Chicago, Illinois, November 19, 2012.

8 Donna Haraway, "A Cyborg Manifesto: Science, Technology, and Socialist-Feminism in the Late Twentieth Century," in *Simians, Cyborgs, and Women: The Reinvention of Nature* (New York: Routledge, 1991), 151.

the text they construct "should not necessarily be equated with any actual historical reality" nor should the stemma they construct be seen as "how it really was."[9] Like the cyborg, the *Ausgangstext* as constructed by these scholars is ahistorical, having no descendants and hailing from no real, originary environment. It represents all extant texts, of every date, compressed and analyzed at one moment of analysis, and this analysis may change from moment to moment, given the iterative, self-correcting (and thus cybernetic) process of the method.

At first glance, this approach to texts seems quite postmodern.[10] It makes no historical claims, and the stability of the baseline text is not a given. But the biological metaphor that contextualizes the method defines text and meaning in positivist, empirical terms. The text as a "carrier" of meaning is analogous to a human with a genetic code that might or might not be expressed in outward appearance, whereas *meaning* is the actual genetic code that lies beneath and is definite and stable, no matter how outward appearances are disfigured through injury or disease. Meaning, then, precedes the text, and the act of interpretation *should* be the "correct" reading of that genetic code. The human mind reading the text should grasp the underlying meaning, or, to put it in terms of DNA, it should process the code (genotype) correctly in order to generate its correct expression (phenotype).

Ironically, the foundation of both the living-text metaphor of narrative textual critics and the text-as-carrier metaphor of CBGM posits an

---

9 Gerd Mink, "The Coherence-Based Genealogical Method (CBGM)," website of the University of Münster Institute for New Testament Textual Research, downloadable presentation (Münster: University of Münster Institute for New Testament Textual Research, 2009), slides 17 and 35, respectively, accessed February 22, 2014, http://www.uni-muenster.de/INTF/cbgm_presentation/download.html.

10 Others have remarked on how the multiple identities of the New Testament seem to mark textual criticism as a field that unexpectedly shares certain points of discussion with critical theory. See Stephen D. Moore and Yvonne Sherwood, "Biblical Studies 'after' Theory: Onwards Towards the Past, Part One: After 'after Theory,' and Other Apocalyptic Conceits," *Biblical Interpretation* 18 (2001): 19: "For textual criticism, that driest and dustiest of biblical disciplines and, one might imagine, farthest removed from the exotic excesses of Theory, the text is a para-postructuralist object. Incurably infected with self-division, it is 'at least dual.' It is, in fact, myriad . . . Though certain of the premises of textual criticism are on a head-on collision course with Theory (not least around the dream of accessing origin and intention), 'textuality,' that Theoretical concept *par excellence*, has certain uncanny affinities with textual criticism." Quoted in Michael W. Holmes, "From 'Original Text' to 'Initial Text': The Traditional Goal of New Testament Textual Criticism in Contemporary Discussion," in *Text of the New Testament in Contemporary Research*, ed. Ehrman and Holmes, 680.

autonomous text. So, in the end, do Bengel's classification and Lachmann's genealogy. In all these iterations of biological metaphors in textual criticism, the text is understood as a living organism, whether with a skin color and disposition, when it is classified, or as a promiscuous, indiscriminate lover that bears corrupted offspring, when its genealogy is traced. These various conceptions of the text as biological being, born of various assumptions and exigencies, demonstrate our age-old tendency to equate our texts with our selves, in order that we may believe our meanings and intentions have a comforting stability and continue to live on as our immortal selves.

None of this, however, is immediately apparent in the biological discourse of textual criticism because it belongs to the privileged discourse of science, that language which promises objectivity, certainty, and reality beyond human desires and fears. And apart from its utility in providing metaphors and methodologies, this scientific discourse has given textual critics a way to speak of their work as if they could maintain a clinical distance from their object of study. By using the discourse of science, textual critics can stand their ground firmly in lower criticism and bracket out the frustrating tangles of literary theory that supposedly belong only to higher criticism. And as the field of philology is threatened by poststructuralist ideas of language, textual critics have turned to scientific discourse as a defense and justification of their work: textual criticism is a *science*, it serves as the foundation, the very reality upon which all interpretive work is based, and, as such, its existence cannot be eradicated because it will always be necessary. As Ehrman states, "you can't know what an author meant if you don't know what he or she said."[11] Therefore, the "scientific" work of textual criticism must first be done before interpretive work begins.

But science itself shifts and changes its premises and goals, so that its very identity as a category of knowledge or experience is unstable. By following the trends of science, textual critics have not anchored their work on an unchanging foundation but have only joined their fortunes with scholars of another shifting field. Furthermore, the discursive formation of philology as a science has masked the interpretive assumptions and strategies of textual critics, which they themselves would dismiss as alien

11 Bart D. Ehrman, "Text and Interpretation: The Exegetical Significance of the 'Original' Text," in *Studies in the Textual Criticism of the New Testament*, New Testament Tools and Studies 33 (Leiden: Brill, 2006), 309.

to their field. The clinical distance textual critics maintain from their work is a fiction.

A return to the metaphor of the tree here serves to demonstrate this point. The tree of textual transmission in New Testament textual criticism almost invariably ends with the advent of printed editions. Of course, given that textual critics have for the most part sought to establish the "original text," this seems an obvious choice: editions are composites of numerous texts and belong to a different era of textual transmission since they were created with modern goals and methodologies. But in another perspective, textual critics continue the same activity and production as the scribes whose work they study: copying from what materials they have, they create new texts for use by scholars, churches, and the general population. Yet almost no textual critic places their work on the *stemma codicum* or considers that they themselves are part of the textual transmission of the New Testament.[12] Consider the following summary of editions in Metzger's *The Text of the New Testament*:

> No one knows how many separate editions of the Greek New Testament have come from the press since 1514, but it is undoubtedly a very great number indeed. Eduard Reuss of Strasbourg, who published a description of editions issued up to 1869, was able to enumerate 584 separate ones. If one adds the re-editions, the variant editions, and some doubtful editions that Reuss mentions in part, the number amounts to 853. Furthermore, since Reuss's list is not complete for the more recent period that he covers and many editions have appeared since he published his volume, it is altogether probable that the 1,000 mark was passed early in the twentieth century.[13]

---

12 One major exception is found in Parker, *Living Text*, 212–3, who considers electronic texts as part of the tradition of the "living text": "it is the tradition in some churches that, when the Gospel has been read at the eucharist, the book is lifted up with the words 'This is the Gospel of Christ.' Whether the book is manuscript or printed, the act is a declaration that it is this particular volume which is the Gospel. For the early Christian communities, the written Gospel existed as *their* Gospel book. Third, this book has not been concerned with a process that is now ended. Scholars continue to discover fresh materials and to revise the text of the Gospels; translators provide fresh renderings, commentators new interpretations, and theologians new ideas. We have seen how much the printed text has changed since 1516, and have wondered what changes will be brought by the electronic text. Today's influences on the text of the Gospel are not those of the early centuries, but they may prove equally powerful. The text of the Gospels remain a living text."

13 Bruce M. Metzger and Bart D. Ehrman, *The Text of the New Testament: Its Transmission, Corruption, and Restoration*, 4th ed. (New York: Oxford University Press, 2005), 194.

It seems that textual critics have been as prolific—if not more so, considering the rate of production—as the scribes. But these editions are not included by Metzger (nor would they be by most other textual critics) in the section "Important Witnesses to the Text of the New Testament." They may be termed "reconstructions," "contributions," or "efforts," but they are not "witnesses." In Metzger's organization of the history of the New Testament text, they belong firmly in the era of "Scientific Textual Criticism"[14] and, as such, do not serve as evidence or belong in the stream of tradition.

But are not these editions also witnesses to the text of the New Testament? Do they not speak to the needs and desires of readers and communities of the text? Do they not represent the latest stage of textual transmission? Textual critics may understand themselves as historians or as scientists, looking to reconstruct a past or a historical narrative, but they also form a part of the history of the text themselves and as such are not "above" the object of study but a part of it. And, ironically, if understood in the terms of their own analogies, they are creating at alarming rates the most contaminated texts of the New Testament of all time!

The case of the *Textus Receptus* also shows certain blind spots of modern textual criticism. This particular form of the text, and its English counterpart, the King James Version, has been derided as a monstrous prodigy by traditional textual critics (Hort's descriptions of it as "villainous" and "vile" come to mind).[15] Alternatively, because of its unchanging yet tenacious nature, it has become a zombie-text—an edition that really should be dead, given its "unscientific" origins, but which continues to shuffle along, preying on susceptible brains. Yet however unpalatable the religious fundamentalism that continues to pump "life" into this form of the text may be, the continued use of the *Textus Receptus* and its defense by a number of Christian scholars indicate that it, too, is part of the textual tradition, not outside of it as an unfortunate anomaly. Its continued use as the "true" text, regardless of whether this claim adheres to "scientific" textual criticism, represents one stream of Christian tradition with its own anxieties and desires.

---

14 The chapter under which the history of editions (post-*Textus Receptus*) is titled "The Modern Critical Period," and the first subtitle is "The Beginnings of Scientific Textual Criticism of the New Testament." Metzger and Ehrman, *Text of the New Testament*, 165.

15 Fenton John Anthony Hort, *Life and Letters of Fenton John Anthony Hort*, ed. Arthur Fenton Hort (London: Macmillan, 1896), 211, in a letter to the Rev. John Ellerton, December 29–30, 1851.

Thus, recent waves of attacks against textual criticism in defense of the *Textus Receptus* are not "an indictment of [the] discipline" that show "how much progress [textual critics] have made in textual criticism in the twentieth century" (i.e., very little).[16] It is not as though the "science" of modern New Testament textual criticism will ever find convincing enough "proof" to permanently dismantle the *Textus Receptus* and convert its defenders. That text is, as much as Codex Bezae or the Oxyrhyncus papyri were, the sacred tradition of a particular community. As much as the "overthrow of the *Textus Receptus*" by Lachmann and Westcott and Hort is discussed in histories of textual criticism, "scientific" textual criticism has achieved no such thing.

Furthermore, this "science" has not progressed (as is the hope expressed by Epp above), building block by building block, toward a final clarity or truth. This understanding of science reflects a Hegelian conception of human history as a progressive and teleological evolution. What the preceding chapters have shown, however, is that goals and methods change, sometimes dramatically, and that there is no independent observer who hovers above the episteme or discourse of the time. Despite their vastly different viewpoints, both the proponents of the *Textus Receptus* and textual critics belong within the tradition of the New Testament, not outside of it.

The discourse of science has created the illusion that textual critics work as scientists, distanced observers of phenomena, whose experiments do not involve the theorizing of interpretive work. The greatest disservice of the use of biological metaphors and methodologies to philologists is its discursive formation of the field as a science, thereby curtailing an exchange of ideas between lower and higher criticism. New Testament textual criticism engages issues of authority and intention, reading and language, but has done so mostly without recourse to the conversations of literary theorists.

It is not the implicit suggestion of this discussion that textual criticism should disengage from scientific discourse or that such a move is entirely possible. However, this deconstruction of the history of New Testament textual criticism reveals the use of scientific discourse as the main force in the production of truth for the field in order to illuminate the

16 Eldon J. Epp, "The Twentieth Century Interlude in New Testament Textual Criticism," the W. H. P. Hatch Memorial Lecture, Society of Biblical Literature, the Palmer House, Chicago, Illinois, November 11, 1973; reprinted in *Perspectives on New Testament Textual Criticism: Collected Essays, 1962–2004* (Atlanta: Society of Biblical Literature, 2005), 85.

boundaries and limitations of that discourse. Conversely, these boundaries and limitations present the areas of greatest opportunity for the field to adopt new methods and metaphors (for we are never free of discourse).

I suggest that the first step toward expanding the epistemological framework of textual criticism would be to dismantle the illusion of the Great Divide of biblical studies and put to rest distinctions between lower and higher criticism that claim the former is historical while the latter is interpretive. Clearly, discourse, episteme, social location, and personal exigencies hold powerful sway not only in scholarly work that deals explicitly with how these interact with the subject at hand but also in fields the very language of which denies the influence of these elements. The second step would be to theorize the field by considering relevant to textual criticism those schools of critical theory generally thought to endanger, malign, or otherwise disregard the tenets of philology. Counterintuitive as this may seem, it is within the exchange of ideas between such disparate-seeming fields that the most productive questions—those addressing the greatest theoretical weaknesses and anxieties—are posed by each to the other.

Poststructuralism and its unmooring of texts from fixed meanings and even from systems of meanings seems a promising place to dive in. In fact, textual critics have begun, intentionally or not, to engage central issues and questions of poststructuralist thought. For example, in his essay "Manuscripts of the Gospels in the Electronic Age," David Parker revisits Luther's characterization of the written Gospels as a necessary evil that is secondary to the preached gospel.[17] The quotation Parker uses comes from Luther's Sermon on the Feast of the Epiphany:

> Man's need to write books is a great injury; it is a violation of the Spirit compelled by necessity, and is not the way of the New Testament. Therefore even Christ himself did not write down his teaching, as Moses did his, but taught orally and likewise commanded it to be continued orally and gave no command to write it. Likewise, the apostle also wrote little, and not all of them did that. . . . For before they wrote they first preached to the people and converted them by the living voice, which was also their proper apostolic and New Testament work.[18]

---

17 David C. Parker, "Manuscripts of the Gospels in the Electronic Age," *Restoration Quarterly* 42 (2000): 221–9.

18 Quoted in Parker, "Manuscripts," 226.

According to Luther, resorting to written works came as a defense against competing traditions and to establish ecclesial authority. The resulting written texts of the Gospels were then subject to all the dangers of history: scribal error, emendation, injury, or total destruction. Those that survived are the texts available to Christians today, and as historical entities form Christian traditions by their very diversity.

According to Parker, the role of scripture-as-tradition should be twofold. On the one hand, the written texts "should be viewed as a companion to the oral tradition, a multiplicity of *aides memoires* to the living Christian community. It is historically still the closest to the historical Jesus that we have."[19] In other words, the written texts are a guide back to an "original." On the other hand, in preaching, the multiplicity of the text should never be ignored, lest scripture be made into one authoritative straitjacket: "It is not the proclamation of a single timeless word . . . The abandonment of the single authoritative text is not a loss, but positively a gain for the preacher."[20]

Two observations should be made at this point. The first is that Parker, like Luther, privileges the spoken word over the written. The written texts are *aides memoires*, "a companion," as close as we can get to the historical Jesus—and his spoken words—but not Jesus himself. Of course, Parker values the multiple texts in and of themselves, but these he also prefers to redefine as spoken rather than written, to emphasize their "free" and "living" nature:

> in its particular nature as a free text, the gospels do have a specific tradition. . . . Each manuscript is unique and in its uniqueness exists as a particular representative of the church's life in its own time and place. That is, it functions more as a *sermon* than as a piece of literature.[21]

The privileging of speech over writing in the history of Western thought was thoroughly deconstructed by Derrida in *De la grammatologie*, published in 1967.[22] The elements at play in contrasting speech and writing, as described by Derrida, are precisely those most fraught in textual criticism: authority, presence, intention, and origin. The spoken word is full of presence,

---

19 Ibid., 225.

20 Ibid., 226.

21 Ibid., 228. Emphasis added.

22 Jacques Derrida, *De la grammatologie* (Paris: Les Éditions de Minuit, 1967). The English translation was published in 1976: Jacques Derrida, *On Grammatology*, trans. Gayatri Chakravorty Spivak (Baltimore, MD: Johns Hopkins University Press, 1976).

while the written implies absence; in Platonic terms, speech is the original, the true form, while text is a copy, a shadow. With the presence and immediate truth of speech, authority and authenticity are guaranteed, whereas with text, these are uncertain. In terms of New Testament textual criticism, the extant texts are only a supplement to the *kerygma* of the Christian gospel; they are literally copies, *aides memoires* to the presence of the historical Jesus. In the search for the "original text," at least as it was originally conceived, the shadowy variants were to lead back to an absolute original and from there to the absolute presence of Jesus, the Word made Incarnate.[23]

This leads to the second observation of Parker's essay: he suggests two basic perspectives from which to view the extant texts of the Gospels. The first, as quoted above, is to view them as a way back to the historical Jesus. The second, by contrast, is to view all of them as forming a pluralistic, sometimes contradictory tradition. These two directions represent an underlying binary opposition at work in New Testament textual criticism today that reflects the dichotomies deconstructed by Derrida, those of speech|writing, original|supplement, presence|absence. The major dichotomy of New Testament textual criticism consists of the aims of narrative textual criticism as versus traditional textual criticism. From this dichotomy can be deduced the numerous other binary oppositions by which the current epistemology of textual criticism functions. They might graphically be represented by Table 5.1.

Table 5.1. The binary oppositions of New Testament textual criticism.

TRADITIONAL TEXTUAL CRITICISM	NARRATIVE TEXTUAL CRITICISM
THE ORIGINAL TEXT	VARIANT TEXTS
ANCESTOR	DESCENDANTS
CLASSIFICATION	EVOLUTION
BASELINE TEXT	APPARATUS
PURITY	DIVERSITY
AUGSGANGSTEXT	LIVING TEXT
SCRIBAL ERROR	SCRIBAL INTERPRETATION
COHERENCE	DIVERSITY

23 This is the analogy Luther makes in the Sermon on the Feast of the Epiphany as summarized by Parker: the text was the embodiment of the preached word just as Christ was the embodiment of God. It is important to note that Parker ultimately rejects this analogy, since there are multiple texts, whereas the Incarnation is unique. Parker, "Manuscripts," 228.

This is not to imply some antagonistic clash between the two sides of these binaries,[24] but to highlight the dichotomies constantly at work in the current discursive form of the field. Breaking down these binaries, that is, deconstructing them, removes the necessity of always privileging one over and against the other and allows scholars greater freedom in the type of questions they ask and the methods they can explore.

What are some of the possibilities, then, if the oppositional binaries of textual criticism are set aside—and how can they be set aside? Certain questions that textual critics are now posing reveal glimpses of unexplored avenues, but the ever-present dichotomy brings the scholar back to well-trodden ways. Two examples serve here to demonstrate the restrictive power of the binary and theoretical questions that can further expand the field. The first is Michael W. Holmes's essay, "From 'Original Text' to 'Initial Text,'" the inclusion of which in the second edition of *The Text of the New Testament in Contemporary Research* signifies a shift in the field toward greater openness to theoretical issues.[25] In tracing the changes in the definition of the "original text" and the adoption of the term *Ausgangstext* or "initial text" in its place, Holmes finally arrives at the fundamental question rarely before discussed in New Testament textual criticism: what *is* text?

> Traditionally, and not without good reasons, the "work" has been understood as an authorial product, an intellectual construct, and the "document" as a specific (and usually imperfect) material representation of it. But . . . some critics argue instead that the work is constituted by the reader's appropriation of it, and/or that it has no existence apart from the material copies (documents) through which it is mediated to the reader. In short, the question of what a "text" (in the broad sense of the term) is has become a matter of substantive discussion.[26]

In presenting the final question "what is text?" Holmes has already anticipated that the answers are binary: either one adopts the traditional view, a

---

24 Indeed, it would be difficult to categorize certain scholars and their work as strictly on one side of the binary or the other.

25 Holmes, "From 'Original Text' to 'Initial Text,'" 637–88.

26 Ibid., 669.

Platonic one, that there is the abstract form of the text and then the physical manifestation of it as a manuscript, or one believes that the text only ever exists as the interpretation of each individual reader. Holmes goes on to discuss Parker's theory that "Scripture *is* tradition" and his argument that the Gospels are *not* "the kinds of texts that have originals." At the end of the discussion, Holmes disagrees with Parker: instead of the Gospels as present only in multiple identities, there is "what one may term microlevel fluidity and macrolevel stability."[27] Again and again, Holmes returns to the binaries of New Testament textual criticism: stability|fluidity, authorial text|individual interpretation, unified|fragmented.

The second example takes us slightly farther. Peter Robinson, of The Canterbury Tales Project,[28] asks the same question, "what is text?" in the intriguingly titled article, "What Text Really Is Not, and Why Editors Have to Learn to Swim."[29] Robinson approaches the question from the perspective of one who works extensively with transcription and machine-readable texts, so that this fundamental question reaches down to the level of individual letters—for example, is an "i" always an "i" or is it interpretable? Robinson's answer to the question at first seems squarely on the side of a more postmodern viewpoint, namely that the text is re-created anew by each interpretation by an individual:

> "What is text, really?". . . text, really, is not. Text does not exist outside the meanings we create: and these meanings are all the text we will ever know. Each text we create is a system of signs: we share these signs with others, and thus the individual text of each of us becomes part of our communal experience. Text is not a tree, which will live whether we perceive it or not. For, text without human perception is just marks on paper, or sounds in the air. These marks and sounds only become text when we find meaning in them.[30]

---

27 Ibid., 674.

28 See Chapter 4.

29 Peter Robinson, "What Text Really Is Not, and Why Editors Have to Learn to Swim," *Literary and Linguistic Computing* 24 (2009): 41–52.

30 Ibid, 45. Robinson, in posing the question and answer, paraphrases from two different articles, respectively: Steven J. DeRose et al., "What Is Text, Really?" *Journal of Computing in Higher Education* 1.2 (1990): 3–26 and Alois Pichler, "Transcriptions, Texts and Interpretations," in *Culture and Value: Philosophy and the Cultural Sciences*, ed. Kjell Johannessen and Tore Nordenstam (Vienna: Austrian Ludwig Wittgenstein Society, 1995), 690.

Robinson labels this position "anti-realist,"[31] and seems to adopt it for most of the article. But toward the latter half of the article, Robinson admits that he believes that a text does, indeed, "speak" to its readers: "the text does have an independent existence; it is saying something; we are not just "constructing" an artefact for the use of our readers, but we are trying to interpret an utterance."[32]

Why does Robinson contradict or at least mediate his first statement regarding "text"? He does so because his main concern, as an editor of *The Canterbury Tales,* is what kind of edition is to be produced and how it will serve its readers. For Robinson, the chief concern is not whether a realist or an anti-realist position is "correct" but rather that they present a "Scylla and Charybdis in textual criticism.[33] An anti-realist position runs the risk of producing a bland edition that helps no reader by trying to serve every one; a "realist" position, on the other hand, could present a narrowly focused edition posing as an authoritative and restrictive text.[34] Again the binary opposition of textual criticism appears.

However, what takes Robinson a bit further than the dichotomy and sets him slightly apart from Holmes is that he asks *why* and *who,* not simply *how* and *what*: why are editions created? for whom are they created? why does one do the work of philology? By contrast, Holmes asks—and I would argue most textual critics do the same—mainly *what* is the goal of New Testament textual criticism and *how* does one achieve it? Holmes argues that the *Ausgangstext* can and should "become accepted as the basic goal of the discipline," because he believes it has the potential to satisfy both sides of the dichotomy.[35] Although he intentionally engages with the epistemological reaches of the field and argues for its expansion, Holmes's concern, in the end, is defining *what* textual criticism should do, not *why* textual criticism is.

For Robinson, the *why* of textual criticism lies primarily with *who*: textual criticism exists to present an edition to readers that is usable

31 Or rather, he adopts the term from Allen Renear, who used it in a paper, "Theory and Metatheory in the Development of Text Encoding," published in an online discussion group. See Robinson, "What Text Really Is Not," 46.

32 Ibid., 47.

33 Ibid.

34 Ibid., 47–8.

35 Holmes, "What Text Is Being Edited?," 681.

and pleasurable and, most importantly, helps readers construct the text for themselves:

> We want readers to understand what it is they are reading, so that they can make texts for themselves with new intelligence. If they do this, then our text will be outdated: and frankly, it will not matter then if people can no longer read our text, and I will not care if they cannot.[36]

With admirable humility regarding the transience of editions, Robinson simultaneously embraces, to borrow from Roland Barthes, the readable text—"what it is they are reading"—and the writable one—"so they can make texts for themselves with new intelligence."[37] With the reader ever in mind, Robinson presents one possibility by which the binary can be subverted: the possibilities of digital texts.

With digital texts, particularly those developed with the questions of *why* and *who* ever at the forefront of their design, capabilities, and manipulability, textual critics can avoid impaling themselves either on one horn or the other of the textual dilemma. Digital texts, especially those of the New Testament, the traditions of which comprise thousands of individual readings, can be represented in fluid ways, shifting from one baseline text to another, allowing focus on one variant or another, and allowing the *reader* to decide *what* to read and *how* to read.

The possibilities of electronic forms of text has not escaped New Testament textual critics,[38] and the twenty-eighth edition of the Nestle-Aland is now available in digital format, although currently it offers not much more manipulability than a printed copy. The generating engine behind the Nestle-Aland edition, CBGM, also presents several possibilities. Because it functions as an application and requires human decisions regarding the direction of "textual flow" in the stemma it creates, there are multiple methodological possibilities for the user constructing the stemma. Of

36 Robinson, "What Text Really Is Not," 50.

37 See *texte lisible* and *texte scriptible* in Roland Barthes, *S/Z: An Essay*, trans. Richard Miller (New York: Farrar, Straus, Giroux, 1974).

38 See Epp's suggestions for possibilities in Eldon J. Epp, "It's All about Variants: A Variant-Conscious Approach to New Testament Textual Criticism," *Harvard Theological Review* 100 (2007): 275–308. See also David C. Parker, "The Text of the New Testament and Computers: The International Greek New Testament Project," *Literary and Linguistic Computing* 15 (2000): 27–41 and id., "Through a Screen Darkly: Digital Texts and the New Testament," *Journal for the Study of the New Testament* 25 (2003): 395–411.

course, CBGM was constructed with the *Ausgangstext* in mind as the goal, so its capabilities are restricted toward that end, but methods and software like those used in CBGM also present possibilities for interacting with the extant texts of the New Testament in new ways.

In anticipation of what new possibilities the digital text may present in the future for textual criticism and its raison d'etre, a new metaphor is needed. One overarching problem of biological metaphors is that they present the text as living and therefore, to an extent, autonomous, whether the text is degenerating, mutating, or adapting. This autonomy, no matter how faintly implied through the metaphors employed, privileges the concept that a text speaks or gives utterance over and against the text that exists only through the interpretive act of the reader. For this reason, a couple of alternative metaphors, though having theoretical potential, are less than desirable because they perpetuate the dichotomy of either the readable or the writable text.

The first alternative is to view the text as rhizomatic and the textual tradition as formed by growing rhizomes rather than a tree. The replacement of the tree with rhizomes is proposed by Gilles Deleuze and Félix Guattari in *A Thousand Plateaus*. Rhizomes are subterranean plant stems that root and send out shoots, and grow unpredictably and out of sight.[39] Rhizomes are agenealogical, they have no hierarchy, do not bifurcate nor follow along a central trunk. In evaluating the field of linguistics, which share the tree and biology metaphors with textual criticism,[40] Deleuze and Guattari offer this comparison, quoted here at length because of the affinities between linguistics and philology:

> A rhizome ceaselessly establishes connections between semiotic chains, organizations of power, and circumstances relative to the arts, sciences, and social struggles. A semiotic chain is like a tuber agglomerating very diverse acts, not only linguistic, but also perceptive, mimetic, gestural, and cognitive: there is no language in itself, nor are there any linguistic universals, only a throng of dialects, patois, slangs, and specialized languages. There is no ideal speaker-listener, any more than there is a homogeneous linguistic community. . . . There is no mother tongue, only a power takeover by

39 Deleuze and Guattari, *A Thousand Plateaus*, esp. 3–25.

40 See Henry Hoenigswald and Linda Weiner, eds., *Biological Metaphor and Cladistic Classification: An Interdisciplinary Approach* (London: Frances Pinter, 1987).

> a dominant language within a political multiplicity. Language stabilizes around a parish, a bishopric, a capital. It forms a bulb. It evolves by subterranean stems and flows, along river valleys or train tracks; it spreads like a patch of oil. It is always possible to break a language down into internal structural elements, an undertaking not fundamentally different from a search for roots. There is always something genealogical about a tree. It is not a method for the people. A method of the rhizome type, on the contrary, can analyze language only by decentering it onto other dimensions and other registers. A language is never closed upon itself, except as a function of impotence.[41]

If "language" were replaced with "texts" in the quotation, it seems that rhizomes offer a powerful metaphor by which to conceptualize the history of New Testament texts. For example: "There is no mother text, only a power takeover by a dominant text within a political multiplicity. Texts stabilize around a parish, a bishopric, a capital. . . . A text is never closed upon itself, except as a function of impotence."

But rhizomes present a similar problem to the tree because they live and grow, implying an unknown, innate, living force within the rhizomatic structure, be it language or textual transmission, that operates autonomously. The rhizome, in this way no different from the tree, presents again a "natural" metaphor, in which analogies are built along organic lines that are predictable once the logic of rhizomatic growth is understood. The rhizome metaphor also only emphasizes multiplicity, one side of the text-critical binary, which paradoxically cannot serve the multiple aims of textual criticism.

A second alternative metaphor comes from postcolonial theory, that of hybridity as conceptualized by Homi Bhabha in *The Location of Culture*.[42] Hybridity in this understanding is not a melding or blending of disparate parts but is interstitial and without fixed identifications and hierarchy. It is an attempt to break down the binary of the colonizer-colonized and is to be understood as "a difference 'within,' a subject that inhabits the rim of an 'in-between' reality."[43] The colonial subject never rests upon one fixed identity but shifts continually from one to another and plays upon difference constantly.

41 Ibid., 7–8.

42 Homi K. Bhabha, *The Location of Culture* (New York: Routledge, 1994).

43 Ibid., 13.

This presents a dynamic way to understand the New Testament, a liminal, shifting, and hybrid entity, containing myriad "differences within." What makes this an even more powerful metaphor is that it is also historical, contextualized by historical narrative, no matter how shifting that narrative may be. Furthermore, other elements of postcolonial theory also seem particularly (and surprisingly) well-suited for New Testament textual criticism. For example, the concept of "mimicry" as not-quite, not-right repetition of colonial discourse, which both repeats and subverts it, may serve as a way to describe variant readings and/or scribal alterations. Additionally, the questions that postcolonial theory asks are pertinent to interpreting the history of New Testament textual transmission: what discourse is privileged? what is a subaltern text? where is the location of power?

It furthermore seems fitting to use postcolonial theory to deconstruct and reconceptualize a field in which the major concepts of classification, genealogy, and corruption are rooted in a colonial, imperialist past. Nevertheless, postcolonial theory, while promising in certain aspects, also presents the problem of configuring the text as Subject, in this case the colonial subject. The concept of hybridity is also a slippery one in which it is incredibly easy to fall back on reliable dichotomies, no matter how emphatically one claims that they are only "heuristic." Besides all this, the term "hybrid" is already in use, outside of its postcolonial meaning, in New Testament textual criticism.

In proposing a new metaphor, I would like to return to the figure of the cyborg, previously discussed in conjunction with the methodology of CBGM. The cyborg is a startling and unlikely figure in the field of New Testament textual criticism, but its jarring presence may just unsettle methods, terms, and delineations enough to shift analysis and understanding in new directions. The theoretical qualities of the cyborg discussed here are loosely inspired by Haraway's concept in the "Cyborg Manifesto," and although (or because) that was written in the interest of deconstructing binaries and essentialisms in feminist theory, its underlying emphases are no less applicable here.[44]

For the discussion of CBGM, the text-as-cyborg seems an apt metaphor because the text, in this case, is a creation of human-machine collaboration. It is also, like a cyborg, a synthesis of all (or close to all) prior knowledge for the creation of an entity wholly new. But the cyborg can serve

44 See Haraway, "A Cyborg Manifesto."

as a synthesis of prior knowledge not only as the text produced by CBGM but also as the figure of the whole textual tradition in New Testament textual criticism, especially as digital texts become dominant. The cyborg's ambiguous, irreal, promising, and threatening character can serve as a synthesis of the prior knowledge of the field while presenting both a deconstruction of restrictive dichotomies and new ways of thinking.

Unlike the tree, the rhizome, or the hybrid, a cyborg is not biologically determined. It is not, therefore, a reflection of our selves as living, reproducing, erotic beings, and so it breaks the Platonic mimesis in the metaphors of text-as-self, text-as-natural organism. The strength of the tree metaphor is always its weakness: texts indeed seem to reproduce and generate genealogies; that is a useful way to visualize the process of transmission. But texts do no such things. Cyborgs are explicitly a reproductive dead-end in and of themselves; they have no erotic life. To use the cyborg metaphor, one must acknowledge that copying, or cloning, always involves *human* agency.

Cyborgs have no genealogy—but they do have an origin and creator, which may or may not matter in its identification, to which it may or may not be faithful. Cyborgs, in our current imagination, present the threat of insurrection, and so they subvert the understanding of authorial intention; but at the same time, cyborgs are programmable and manipulable. In this way, the text-as-cyborg metaphor need not fall on either side of a binary because its nature presents a constant cycle of reconfiguration, release, and subversion.

Cyborgs, as cybernetic organisms, make difficult—but not impossible—the language of gender and race, sex and family, purity and corruption. They are not easily placed in taxonomies. They function atheologically, so that they resist the notions of authenticity, contamination, and fallenness we otherwise are prone to attribute to texts, especially the texts of scripture. Again, cyborgs seem like us, but always point back to us, to human action and participation, in the assignment of category, degeneration, intention, or reproduction.

Finally, to talk of cyborgs is to shift from scientific discourse to science fiction. This at once acknowledges the imagination and beauty of scientific endeavor while valuing the craft of narrative. The unexpected, fantastical figure of the cyborg reminds us that we work within discourse, not above it, and the histories we construct may or may not ever materialize as truth.

APPENDIX I

Excerpt of Interview with Gerd Mink and Klaus Wachtel, Institut für Neutestamentliche Textforschung, Münster, March 10, 2011

[This excerpt begins with the discussion of the possibility of using phylogenetic software and methods in the field of New Testament textual criticism.]

WACHTEL: We made some experiments with this and biological chemistry—there was a project called Stemma, with Christopher Howe at Cambridge, and on the other side, from the philological perspective there was Peter Robinson. They set up this project and tried and did apply some software developed for genealogical research in biology. Yes, it was a very interesting time.

I worked together with a colleague, Matt Spencer, and we published an article together about a well known group of manuscripts that is the group that is behind a precisely datable translation from the early seventh century into Syriac by Thomas of Harqel.[1] So we had this group and performed different methods on it, but one [discovery we made] was that our questions to the material are not the same as those that biologists ask when they look at the evidence.

And there are some fundamental differences: one is we are looking at a kind of, yes, one might call it a degeneration and not an evolution. Of course, the

1 Matthew Spencer, Klaus Wachtel, and Christopher Howe, "The Greek Vorlage of the Syra Harclensis: A Comparative Study on Method in Exploring Textual Genealogy," *TC: A Journal of Biblical Textual Criticism* 7 (2002), accessed May 2, 2013, http://www.reltech.org/TC/v07/SWH2002/index.html.

persons who think that the Majority Text is the original one, they might think that it was an evolution. But, no, not even they because they say what we have in the Majority Text was already there in the beginning.

Lin: But is that just a question of terms?

Wachtel: No, it is a fundamental difference. All these programs that are used by biologists do not input genealogical information of any kind. They start from a tree showing the relationships between nodes, nodes of states that have developed. And they explicitly do not want to think about genealogy in the first place, but they want to see a structure. And we are in a way different because we do not exclude philological assessment of variation units. We have already talked about the local genealogy, to which method this Institute was restricted until [Mink] developed the Coherence-Based Genealogical Method, which takes into account the whole set of variation units in a given writing, that's the big difference from the mere local genealogy. But at each passage of variation we introduce our philological assessment and take that into account.

Mink: But what you get at the end is not a tree of manuscripts; it is a tree of the texts in the manuscripts, and the text is what gives value to a manuscript. So manuscripts themselves [are] parchment, or whatever [they] may be. The text is an abstract thing.

Lin: Well, that's one of the things I wanted to ask as I was looking over your article, "Problems of a Highly Contaminated Tradition."[2] You talk about the manuscript as being the physical carrier of the text, but then you speak of the text being the carrier of meaning, and I wanted to ask you what, then, is "meaning," in your definition? You say that at the level of the character, in the characters on the page we have variation, and those might deform what the meaning is. Are text and meaning interchangeable at some point? Or are they distinct values?

Mink: You may define a text as a chain of letters but you don't read a text as a chain of letters. And a scribe doesn't copy a text as a chain of letters. Maybe if he has a very difficult word that he has to work letter by letter, but as you read a text you see the meaning of the text, the utterance, the sense of a text. And the normal behavior of a copyist is, as we see, not to remake a text or to rewrite but to [produce] the same physical form of the sense of a text. And again and again it happens that he has the sense in mind and copies, and the result is slightly different but the sense is the same.

Lin: So I would ask, then, would "meaning" be closest related to what you would say might be the "authorial intent" as given by the originator of the text?

---

2 Gerd Mink, "Problems of a Highly Contaminated Tradition: The New Testament: Stemmata of Variants as a Source of a Genealogy for Witnesses," in *Studies in Stemmatology II*, ed. Pieter van Reenen, August den Hollander, and Margot van Mulken (Amsterdam: John Benjamins, 2004), 13–86.

Mink: Authorial—I don't know. We don't know which steps were between the authorial text and the text as published the first time. But I think that the simplest hypothesis is that we have to search for the authorial text, maybe with some alterations which cannot be avoided in copying a text. But if there is evidence that this cannot be the authorial text, for instance if you have the hypothesis that two letters have been compiled and in compiling these letters the redactor has felt the need of changing something, then the authorial text—yes, this is a point [where authorial intention is significant]. Textual criticism, in the traditional view, is to look only for the development of the text of the so-called ur-text into variations. And textual criticism has a result that we can say, "This is the oldest form of the text." But textual criticism cannot decide on the authorial text, [that's] a question of exegesis. . . . But if we have a text and ask what may be the older text, arguments from exegesis are indispensable. For instance, if I know the purpose, the aim of an author and I find a variation which is [counterproductive] with this aim, it is very unlikely to be preferred.

Wachtel: Even if it is the most difficult text. Maybe we should just get this distinction clear, the difference between the initial text and the authorial text.

Lin: Right, the *Ausgangstext* and the "original text," which has been questioned many times. Actually, I was at SBL in Atlanta, and I remember seeing you [Wachtel] there, asking a question of Eldon Epp because he was speaking about, he did this whole presentation on what could the "initial text" mean.[3] Maybe you are tired of the question?

Wachtel: No, not at all, but that was a tantalizing presentation because he went into the etymology of "initial." I mean, it did not help to understand what we meant by "initial text." The definition is clear and simple, as I tried to say at the presentation, too. It is different, it is important to distinguish between the point from which our MS tradition started and what the author wrote in the very beginning of this process. And, as Gerd already said, what lies between the initial text on one hand and the authorial text on the other is viewed (at least to a large extent—we do not exactly know)—but then it is also clear that the initial text must have some relationship with the authorial text because otherwise the exegetes were at a complete loss. What would they interpret if they did not think that there has been an author who wrote this text for a specific reason and with a specific aim and so on? But we do not know at every turn whether we have the original text's authorial wording or whether there has been some change in between the writing of the text and then the beginning of the MS tradition as we have it.

---

3 Eldon J. Epp, "In the Beginning Was the New Testament Text, But Which Text? Consideration of the Terms *Ausgangstext* and 'Initial Text,'" talk given in the New Testament Textual Criticism session "Honoring the Work of Eldon Jay Epp, Octogenarian," at the Society of Biblical Literature Annual Meeting, Atlanta, Georgia, 2010.

Mink: I think working on each writing of the New Testament we must define what, in this special case, the initial text means. And we published the Catholic Letters; our hypothesis, reconstructing the text, was that the authorial texts will be near to what is the initial text. In other parts of the New Testament it may be different. And you know the tendency is to say the text we can reconstruct is of the second half of the second century and so on—or the fourth century—so they have enough room in the second century to play with variant hypotheses, what may have happened in the evolution of the text. And *The Living Text of the Gospels* by David Parker[4] or *The* [*Orthodox*] *Corruption* [*of Scripture*] by Ehrman[5] and so on [are examples of this]. But what is the evidence? What is the evidence? Our experience is that nearly all texts have very close relatives. You cannot see it if you read the Nestle-Aland, but if you have the full evidence you see it. There is one or the other close relative and the best hypothesis is that we had a great continuum of textual tradition. There are exceptions, that's very clear.

Lin: But I wanted to ask about some of the basic principles, the terminology [of the Coherence-Based Genealogical Method], and what you thought about new developments in the field that I think go against those assumptions. For example, we just talked about David Parker and Bart Ehrman, and one of the assumptions [of the Method] is that the scribes do not intend to change the text and they work with as few texts as possible—the rule of parsimony, which you use in the presentation.[6] So what is your critique or understanding of something like *The Orthodox Corruption of Scripture* when [Ehrman's] definitely arguing that, well, more or less, there's scribal intention to change [due to] Christological or theological debates?

Wachtel: One principle that Ehrman refers to is as old as Jerome. He says that scribes often do not write what they see in their exemplar but what they understand it to mean, and that's certainly true. But that's not authorial intention [on the part of the scribe], but that's what we find in the manuscript tradition of all writings from antiquity. But when it comes to the intention to change the text to make it conform with orthodox theology then I think he exaggerates this principle into the direction that the scribe becomes a kind of an author, and that's

4 David C. Parker, *The Living Text of the Gospels* (Cambridge: Cambridge University Press, 1997).

5 Bart D. Ehrman, The Orthodox Corruption of Scripture: The Effect of Early Christological Controversies on the Text of the New Testament (Oxford: Oxford University Press, 1993).

6 Gerd Mink, "The Coherence-Based Genealogical Method (CBGM)," website of the University of Münster Institute for New Testament Textual Research, downloadable presentation (Münster: University of Münster Institute for New Testament Textual Research, 2009), accessed February 22, 2014, http://www.uni-muenster.de/INTF/cbgm_presentation/download.html.

too much. I think that the transmission process was influenced by the beliefs of scribes, in the sense of this principle that they might write down what they understand, not what they should copy. But authorial intent is something different and I think the most convincing argument against such a revision according to orthodox principles is that you find them in one place but in the next verse it may be completely different. And that is a point that Ehrman never really deals with. He refers to it and says, "Well, I know there is inconsistency, but it doesn't matter, I explained it elsewhere." But if you look elsewhere you don't find it explained, at least that was my experience. So in part he is certainly right that there is something in existence like orthodox corruption, if one understands it correctly, but intended revision of the text is something completely different. And we see that in the Byzantine text, we find very many passages that need to be conformed with Byzantine orthodox theology, that's for sure, and they were copied without variants in many cases; in other cases, yes we have this.

MINK: And you must see that the proportion of variants which may be concerned is very small and it becomes greater if you look through the right glasses. And in most cases it is clear what caused the variation or you can imagine what may have caused the variation—the context, eye-skipping, and very simple things. And if I asked why did this variant develop, the first answer or the best, the preferable answer is a simple answer, and not that there was great effort in the brain of the scribe.

WACHTEL: No, certainly not on the scribe's side. That's another distinction that is important, the scribe and an editor. I don't think that they in many cases were the same person. There must have been some editorial activity in the transformation, that's for sure, but, poor scribes, usually they had to copy.

LIN: So that brings me to my next question. It seems that in the last ten to twenty years the field has been [moving] towards historical reconstruction using variants, from what I'm reading of David Parker, Bart Ehrman, Eldon Epp, what they call "narrative textual criticism." And I wanted to ask you both what do you think is properly in the field of textual criticism and at what point—how would you define that, because it seems that we are moving beyond strictly traditional textual criticism. It almost seems, when I read *The Living Text of the Gospels* or his other *Introduction to New Testament Manuscripts and Their Texts*,[7] David Parker seems to be trying to *rescue* the field of textual criticism to a certain extent because he sees—or I see a fear that [others will say], "Oh, this field is finished and we've completed the task," and he wants—many want to say, "Oh, but we're going to do this instead." And I don't know—I mean, there's obviously plenty of work to be done for the *Editio Critica Maior*.

---

7 David C. Parker, *An Introduction to New Testament Manuscripts and Their Texts* (Cambridge: Cambridge University Press, 2008).

Wachtel: By the way, I heard Bart Ehrman once put it—I even remember the year, it was 1993 when SBL International was in Münster—he said [that] the textual criticism of the New Testament is nowadays just a mopping-up exercise, so all the basic problems have been solved and there's no reason to apply textual criticism anymore. And he explicitly said that at times. But David Parker certainly would not subscribe to this statement. I know him very well, and, yes, he and his group visited Münster just last week, and we are going to work on the *ECM*. So he is interested in the history of the text, but his book, *The Living Text*, and the other publications, the focus is more on the single witness, on the single manuscript. And you know he wrote a monograph about one single manuscript and in very great detail.[8] So these manuscripts no doubt have stories to tell, and they are very interesting artifacts under many perspectives, art-historical, paleographical, cultural-historical, no doubt about that. But we are working together on the Gospel of John now, and will apply the Coherence Method, so he is, yes, open to both aspects. If he performs something like narrative textual criticism, well, then, that is one aspect of the history of the text that he likes to focus on—under a certain perspective that does not exclude textual criticism.

Mink: No, we cannot speak of the end, we cannot say that textual criticism is going to become superfluous. You can see that the discussion on text types is very heavy in the last years, and it is not clear what the result will be. We formulated our doubts concerning the Catholic Letters—the only text type there is is the Byzantine text and the others [that] are not Byzantine, and there is a great field and great differences between the various strands of tradition. As Klaus said, it is very important that people deal with individual manuscripts with readings and to learn more about the way of copying, what does the scribe—perhaps you have seen Evangelical Text Criticism,[9] there was a discussion on the blog whether we should speak of scribes or copyists, yes? Well, in contrary to those which deal with individual manuscripts, our task in this institute is to edit texts, to make an edition, and our edition has a primary line and we have certain claims concerning this primary line. It was not disputable for us to take a text, say the *Textus Receptus*, or something like that, or to proceed like the edition of Luke, the International Greek New Testament Project edition, where the ideal was to present the material and what primary line there is—

Lin: You choose yourself.

Mink: Yes.

Wachtel: They thought that was just the preliminary task and that the edition proper would follow, which now has been given up.

---

8 Id., *Codex Bezae: An Early Christian Manuscript and Its Text* (Cambridge: Cambridge University Press, 1992).

9 Peter M. Head and Tommy Wasserman, eds., *Evangelical Text Criticism*, accessed Feb. 26, 2014, http://evangelicaltextualcriticism.blogspot.com/.

Mink: Yes, and if you have a primary line, and we have to claim that this primary line is reconstructed on the—based on all that we know about the textual history—but what do we know about the textual history? It is the first time now that we have such an amount of material and we used computers.

Lin: You don't have to have students collating!

Wachtel: We *have* students—well, they transcribe.

Mink: But the analysis of all the material requires computers, and to construct a global stemma only of 124 witnesses in the Catholic Letters is impossible working with desktop computers. And the tests were possible only with witnesses which have only a few potential ancestors. But if witnesses have more than twenty potential ancestors you must test all possibilities which may occur, and it is only possible with a high performance computer which we have now in our university since the beginning of this year. And I hope that we can start with producing the stemmata.

Lin: I wanted to ask—in the presentation it was very clear that the CBGM is not in itself a method but a meta-method in which you make your own decision, [using] whatever philological foundation you want to put in. So what would you characterize as the philological foundation for what happens with the *ECM*?

Wachtel: The philological foundation? The traditional methods that have been known—

Lin: The canons that have been passed down?

Wachtel: Yes, even, say, that Jerome already knew about, the basic principles. Well, that has been developed further and nowadays the first modern textual critic we would perhaps base the tradition in, that we are still in, is Lachmann, I think, who first in modern times systematized the principles, although he was, he worked differently than us. His aim was not in the first place to reconstruct the original text.

Lin: But the text of the fourth century, right?

Wachtel: Yes, the fourth century's text, and he just wanted to present the evidence in an objective way and present it to scholars.

Mink: The canons of textual criticism have a certain ambiguity. Everybody quoting these canons says that you cannot apply them without caveats, and in some cases the contrary of what the canon says may apply, yes? And [there has been this] discussion: have texts the tendency to become longer or shorter? I think everybody working with texts has imagination of what may be the truth, but he doesn't know actually what the truth is. And we now have a great amount of variations, only in the Catholic Letters more than 3,000 passages of variations, a lot of variants, and the variety of how texts develop is greater than imagined before. And we learn about the principles of variations more and more, and, thus, I think the canons of textual criticism will develop too.

Lin: And also based upon, I think, the instinct of a textual critic.

Wachtel: And the knowledge, the more passages you have seen the more you know about variation.

Mink: What you get with CBGM, on the one hand the structure inherent in the tradition, on the other hand you get the structure inherent in your decisions, in your philological assessments, and this way it is a possibility to control myself. And it is very clear if you have a very different starting point, for instance, the Majority Text is the right text, you make decisions every passage [so that the] Majority is the priority text, yes, you will get with the same method a picture where the Byzantine text is in the first place. It is possible. But you cannot say that, for instance, Minuscule 33 has the oldest text form. If you make all decisions in this way it would be complicated without—it is very complicated. The result is very complicated. The suspicion [would be] great that there is something wrong.

Lin: Which is then what I understand the "iterative process" to be, right? If you make these decisions and then you re-examine how the stemmata look, then, if it's unlikely, if there's little coherence, then you would go back and remake those decisions. Is that correct?

Wachtel: Yes. I also revise the local stemmata in the light of the tendencies that emerge in the summarizations of the decisions taken.

Mink: But I'm not really happy with my iterative process because it can be misunderstood in this way—that you see a preliminary result and now you are feeling—

Lin: As if it's manipulation.

Mink: Yes, manipulation. I have written an essay on coherence and coincidence which will be published, I think, this year.[10]

Wachtel: I hope so, yes.

Mink: I hope this year, in this text-critical series of the SBL.

Wachtel: Yes, that's the contributions to a colloquium on the coherence method that took place in 2008.[11]

Mink: There I make it clear that there is no iteration—there are not ten iterations or something like that. In the case of the Catholic Letters, working was more complex because at the beginning of our edition, letter of James only, we had no data of 2 Peter or Jude, and so on. And we had only a view on James and the relations there, but we don't have the same relations in 1 Peter. Relations between some witnesses may be stable, some have complete other form of texts and other writing, and so on. And it was clear that the picture of the tradition will change if we apply all that we know about the Catholic Letters, and after

---

10 Gerd Mink, "Contamination, Coherence, and Coincidence in Textual Transmission," in *The Textual History of the Greek New Testament: Changing Views in Contemporary Research*, ed. Klaus Wachtel and Michael W. Holmes (Atlanta: Society of Biblical Literature, 2011), 141–216.

11 Colloquium zur Textgeschichte des Neuen Testaments at the Institut für Neutestamentliche Textforschung, Westfälische Wilhelms-Universität Münster, Aug. 3–6, 2008.

having finished Jude, and having gathered the entire material, we can say much more about the relationships of the texts. So, thus, in the beginning our work was very preliminary. And now, after having finished Jude, we have the definite view on what exists. And the local stemmata of variants must be revised based on this total view. And after that it may be possible that we see that we are—we made mistakes in assessing some relationships between variants. The most possible help is that we can see which relations are extremely improbable. The method cannot say the relation must be this one or it [cannot] say this variant and this variant may not [be] related one to the other. But the most important thing is philological assessment.

[The rest of the interview involves interaction with CBGM software and its visual presentation. For that reason, it has been omitted.]

APPENDIX 2

"Marcus Niebuhr Tod"

by Maurice Bowra

I sing of Marcus Niebuhr Tod[1]
Who found the autograph of God.
Carved on a Sinaitic cliff
He found Jehovah's hieroglyph
A casual glance assured the date
As B.C. 1698.
Convinced that no epigraphist
Has entered it on any list,
The worthy Marcus looked around
And found in fragments on the ground
An interesting Hebrew text
Concerning this world and the next.
With scrupulous and loving care
He joined the pieces lying there,
And with a scholar's zeal restored
Words written by the living Lord:
"Go, worship other gods than me:
I feel the need of company.
"And copy bird or beast or fish

1. *New Bats in Old Belfries: Some Loose Tiles*, ed. Henry Hardy and Jennifer Holmes (Oxford: Robert Dugdale, 2005), 3. Marcus Niebuhr Tod (1878–1974) was a renowned ancient historian and epigraphist at Oriel College, University of Oxford.

In wood or stone, if that's your wish.
"Rightly or wrongly, take my name:
To me it's very much the same.
"If toil of week-days you must shirk,
Then keep the Sabbath for your work.
"If what you want's a good long life,
With parents keep continual strife.
"Kill anyone who worries you:
It's easier and safer too.
"And if your neighbour's wife is free,
Of course commit adultery.
"If others have what you have not,
Why not take steps to steal the lot?
"False evidence against a friend
Brings much advantage in the end.
"If someone has what takes your eye,
Ask, and you'll get it by-and-by."
These mystic words did Marcus find,
And laid them carefully to mind,
But not before, upon his knees,
He knelt and took a loving squeeze.[2]
When in the "Epigraphic Year"[3]
The learned saw his work appear,
They pressed the ill-considered view
That here at last was found the true
Editio princeps of the Ten
Commandments given to the men,
Which Moses cast upon the ground.
This text, they claimed, had Marcus found.
But Marcus did not welcome this
Improbable hypothesis,
And thought that it was far too bold
For any prudent man to hold . . .
Perambulating Oriel Quad
I thought of Marcus Niebuhr Tod.

---

2. I.e., a plaster impression of the inscription, used by epigraphists to copy it and potentially to help read the text.

3. The scholarly journal of classical epigraphy, *L'Année Épigraphique*.

Bibliography

Aland, Barbara. "Welche Rolle spielen Textkritik und Textgeschichte für Verständnis des Neuen Testaments? Frühe Leserperspektiven." *New Testament Studies* 52 (2006): 303–18.

Aland, Kurt. "The Significance of the Papyri for Progress in New Testament Research." In *The Bible in Modern Scholarship*, ed. J. Phillip Hyatt, 325–45. Nashville, TN: Abingdon, 1965.

——— "The Twentieth Century Interlude in New Testament Textual Criticism." In *Text and Interpretation: Studies in the New Testament Presented to Matthew Black*, ed. Ernest Best and Robert McLachlan Wilson, 1–14. Cambridge: Cambridge University Press, 1979.

——— "Der neue 'Standard Text' in seinem Verhältnis zu den frühen Papryi und Majuskeln." In *New Testament Criticism: Its Significance for Exegesis: Essays in Honour of Bruce M. Metzger*, ed. Eldon J. Epp and Gordon D. Fee, 257–75. Oxford: Clarendon Press, 1981.

Aland, Kurt, and Barbara Aland, *The Text of the New Testament: An Introduction to the Critical Editions and to the Theory and Practice of Modern Textual Criticism*. 2nd ed. Translated by Erroll F. Rhodes. Grand Rapids, MI: Eerdmans, 1995.

Aland, Barbara, Kurt Aland, Gerd Mink, Holger Strutwolf, and Klaus Wachtel, eds. *1–2 Peter, Novum Testamentum Graecum: Editio Critica Maior*. Stuttgart: Deutsche Bibelgesellschaft, 2000.

Alberts, Bruce, Dennis Bray, Julian Lewis, Martin Raff, Keith Roberts, and James D Watson. "Genetic Recombination." In *Molecular Biology of the Cell*, 3rd ed. New York: Garland Science, 1994. Available at the website of the National Center from Biotechnology Information. Accessed April 29, 2013. http://www.ncbi.nlm.nih.gov/books/NBK28388/.

Alford, Henry. *The Greek New Testament*. Vol. 1. 5th ed. London: Rivingtons, 1863.

Ashton, John. *Studying John: Approaches to the Fourth Gospel*. Oxford: Oxford University Press, 1994.

Barbrook, Adrian C., Christopher J. Howe, Norman Blake, and Peter Robinson. "The Phylogeny of *The Canterbury Tales.*" *Nature* 394 (1998): 839.

Barnes, Jonathan. *Porphyry's Introduction.* Oxford: Oxford University Press, 2003.

Barthes, Roland. *S/Z: An Essay.* Translated by Richard Miller. New York: Farrar, Straus, Giroux, 1974.

Beiko, Robert G., and Nicholas Hamilton. "Phylogenetic Identification of Lateral Genetic Transfer." *BioMedical Central Evolutionary Biology* 6 (2006). Accessed July 27, 2015. http://www.biomedcentral.com/content/pdf/1471-2148-6-15.pdf.

Bengel, Johann Albrecht. "Prodromus Novi Testamenti Graeci recte cauteque adornandi." In *Ioannis Chrysostomi de sacerdotio libri VI.* Stuttgart: Mezler and Erhard, 1725.

——— *'H KAINH ΔIAΘHKH. Novum Testamentaum Graecum.* Tübingen: Cotta, 1734.

——— *Gnomon Novi Testamenti.* Tübingen: Schamm, 1742.

——— *'H KAINH ΔIAΘHKH.* 2nd ed. Edited by Philip David Burke. Tübingen: Cotta, 1763.

Bentley, Richard.. [Phileleutherus Lipsiensis, pseud.] *Remarks upon a Late Discourse of Free-Thinking in a Letter to F.H. D. D.* 8th ed. Cambridge: Bentham, (1713) 1743.

——— *Proposals for Printing a New Edition.* In *The Works of Richard Bentley, D. D.,* vol. 3, ed. Andrew Dyce. London: MacPherson, (1720) 1838.

——— *Bentleii Critica Sacra.* Edited by Arthur Ayres Ellis. Cambridge: Deighton, Bell, 1862.

Bhabha, Homi K. *The Location of Culture.* New York: Routledge, 1994.

Blumenbach, Johann Friedrich. *De generis humani varietate nativa.* 3rd ed. Göttingen: Vandenhoek and Ruprecht, 1795.

Blumenkranz, Bernhard. "Fidélité du scribe." *Revue du Moyen Âge* 8 (1952): 323–6.

Bonnet, Charles. *Contemplation de la Nature, Nouvelle édition, corrigée et considerablement augmentée.* Hamburg: Virchaux, 1782.

Botha, J. Eugene. "New Testament Textual Criticism Is Dead! Long Live New Testament Textual Criticism!" *Hervormde Teologiese Studies* 63 (2007): 561–73.

Bowler, Peter J. *Evolution: The History of an Idea.* Rev. ed. Berkeley: University of California Press, 1989.

——— *The Mendelian Revolution: The Emergence of Hereditarian Concepts in Modern Science and Society.* London: Athlone Press, 1989.

Bowra, Maurice. *New Bats in Old Belfries: Some Loose Tiles.* Edited by Henry Hardy and Jennifer Holmes. Oxford: Robert Dugdale, 2005.

Brace, C. Loring. "One Origin or Many?" In *"Race" Is a Four Letter Word,* 37–43. New York: Oxford University Press, 2005.

Brézé, Jacques de. *Le livre de la chasse du Grand Seneschal de Normandye et Les ditz du bon chien Souillard.* Paris: Aubry, 1858.

Bryant, David, and Vincent Moulton. "NeighborNet: An Agglomerative Method for the Construction of Planar Phylogenetic Networks." In *Algorithms in Bioinformatics: Second International Workshop, WABI,* edited by Roderic Guigó and Dan Gusfield, 375–91. Lecture Notes in Computer Science 2452. Berlin: Springer, 2002.

Buffon, Georges-Louis Leclerc, Comte de. *Histoire naturelle, générale et particulière, avec la description du Cabinet du Roy*. 36 vols. Paris: L'Imprimerie Royale, 1749–88.

Burgon, John William. *The Revision Revised*. London: Clowes, 1883.

Burkitt, Francis C. Review of *The Four Gospels, a Study of Origins* by B. H. Streeter, *Journal of Theological Studies* 26 (1925): 278–94.

Canterbury Tales Project. Accessed July 22, 2015. http://hridigital.shef.ac.uk/canterbury-tales.

Cartlidge, Neil. "*The Canterbury Tales* and Cladistics." *Neuphilologische Mitteilungen* 102 (2001): 135–50.

Coetzee, John Maxwell. "Blood, Flaw, Taint, Degeneration: The Case of Sarah Gertrude Millin." *English Studies in Africa* 23 (1980): 41–58.

Cohen, Philip, ed. *Devils and Angels: Editing and Literary Theory*. Charlottesville, VA: University of Virginia, 1991.

Colwell, Ernest C. "The Genealogical Method: Its Achievements and Its Limitations." *Journal of Biblical Literature* 66(1947): 109–33.

——— "Hort Redivivus: A Plea and a Program." In *Studies in Methodology in Textual Criticism of the New Testament*, 148–71. Leiden: Brill, 1969.

Colwell, Ernest C., and Ernest W. Tune, "Method in Establishing Quantitative Relationships between MS. Text-types." In *Biblical and Patristic Studies: In Memory of Robert Pierce Casey*, ed. J. Neville Birdsall and Robert W. Thomson, 56–62. Freiburg: Herder, 1963.

Crais, Clifton, and Pamela Scully. *Sara Baartman and the Hottentot Venus: A Ghost Story and a Biography*. Princeton: Princeton University Press, 2009.

Darwin, Charles. *On the Origin of Species by Means of Natural Selection*. London: Murray, 1859.

——— *Journal of Researches into the Natural History and Geology of the Countries Visited during the Voyage of HMS Beagle*. 3rd ed. London: Folio Society, (1860) 2003.

——— *Charles Darwin's Letters: A Selection, 1825–1859*. Edited by Frederick Burkhardt. Cambridge: Cambridge University Press, 1996.

Dawkins, Richard. *River Out of Eden: A Darwinian View of Life*. New York: Basic Books, 1995.

——— *The Ancestor's Tale: A Pilgrimage to the Dawn of Evolution*. New York: Houghton Mifflin, 2004.

Deleuze, Gilles, and Félix Guattari. *A Thousand Plateaus: Capitalism and Schizophrenia*. Translated by Brian Massumi. Minneapolis, MN: University of Minnesota Press: 1987.

DeRose, Steven J., David G. Durand, Elli Mylonas, and Allen H. Renear. "What Is Text, Really?" *Journal of Computing in Higher Education* 1.2 (1990): 3–26.

Derrida, Jacques. *On Grammatology*. Translated by Gayatri Chakravorty Spivak. Baltimore, MD: Johns Hopkins University Press, 1976.

Desmond, Adrian J., and James Moore. *Darwin: The Life of a Tormented Evolutionist.* New York: Norton, 1991.

Dobzhansky, Theodosius. *Genetics and the Origin of Species.* New York: Columbia University Press, 1937.

Ehrman, Bart D. *The Orthodox Corruption of Scripture: The Effect of Early Christological Controversies on the Text of the New Testament.* Oxford: Oxford University Press, 2003.

——— "Methodological Developments in the Analysis and Classification of New Testament Documentary Evidence." In *Studies in the Textual Criticism of the New Testament,* 9–32. New Testament Tools and Studies 33. Leiden: Brill, 2006.

———. "Text and Interpretation: The Exegetical Significance of the 'Original' Text," in *Studies in the Textual Criticism of the New Testament,* 307–24. New Testament Tools and Studies 33. Leiden: Brill, 2006.

———. "Text and Transmission: The Historical Significance of the 'Altered' Text." In *Studies in the Textual Criticism of the New Testament,* 325–42. New Testament Tools and Studies 33. Leiden: Brill, 2006.

———. "The Use of Group Profiles for the Classification of New Testament Documentary Evidence." In *Studies in New Testament Textual Criticism,* 33–56. New Testament Tools and Studies 33. Leiden: Brill, 2006.

Ehrman, Bart D., and Michael W. Holmes, eds. *The Text of the New Testament in Contemporary Research: Essays on the Status Quaestionis.* 2nd ed. New Testament Tools, Studies and Documents 42. Leiden: Brill, 2013.

Epp, Eldon J. *The Theological Tendency of Codex Bezae Cantabrigiensis in Acts.* Cambridge: Cambridge University Press, 1966.

———. "The Twentieth Century Interlude in New Testament Textual Criticism," The W. H. P. Hatch Memorial Lecture, Society of Biblical Literature, the Palmer House, Chicago, Illinois, November 11, 1973. Reprinted in *Perspectives on New Testament Textual Criticism: Collected Essays, 1962–2004* (Atlanta: Society of Biblical Literature, 2005), 59–100.

———. "The Eclectic Method: Solution or Symptom?" *Harvard Theological Review* 69 (1976): 211–57. Reprinted in *Perspectives on New Testament Textual Criticism: Collected Essays, 1962–2004* (Atlanta: Society of Biblical Literature, 2005), 125–73.

———. "New Testament Textual Criticism in America: A Requiem for a Discipline." *Journal of Biblical Literature* 98 (1979): 94–8.

———. "A Continuing Interlude in New Testament Textual Criticism?" *Harvard Theological Review* 73 (1980): 131–51.

———. "The Multivalence of the Term 'Original Text' in New Testament Textual Criticism." *Harvard Theological Review* 92 (1999): 245–81.

———. "Introduction: A Half-Century Adventure with New Testament Textual Criticism." In *Perspectives on New Testament Textual Criticism: Collected Essays, 1962–2004,* xxvii–xl. Atlanta: Society of Biblical Literature, 2005.

———. "The Oxyrhynchus New Testament Papryi: 'Not Without Honor Except in Their Hometown'?" In *Perspectives on New Testament Textual Criticism: Collected Essays, 1962–2004*, 743–801. Atlanta: Society of Biblical Literature, 2005.

———. "Toward the Clarification of the Term 'Textual Variant.'" In *Perspectives on New Testament Textual Criticism: Collected Essays, 1962–2004*, 101–24. Atlanta: Society of Biblical Literature, 2005.

———. "It's All about Variants: A Variant-Conscious Approach to New Testament Textual Criticism." *Harvard Theological Review* 100 (2007): 275–308.

———. "In the Beginning Was the New Testament Text, But Which Text? Consideration of the Terms *Ausgangstext* and 'Initial Text.'" Lecture given in the New Testament Textual Criticism session "Honoring the Work of Eldon Jay Epp, Octogenarian," Society of Biblical Literature Annual Meeting in Atlanta, 2010.

———. "Textual Clusters: Their Past and Future in New Testament Textual Criticism." In *The Text of the New Testament in Contemporary Research: Essays on the Status Quaestionis*, edited by Bart D. Ehrman and Michael W. Holmes, 519–77. 2nd ed. New Testament Tools, Studies and Documents 42. Leiden: Brill, 2013.

Erasmus, Desiderius. *Adagiorum Chiliades* Basel, 1538.

Ereshefsky, Marc. *The Poverty of the Linnaean Hierarchy: A Philosophical Study of Biological Taxonomy*. Cambridge: Cambridge University Press, 2001.

Everson, Ted. *The Gene: A Historical Perspective*. Westport, CT: Greenwood, 2007.

Eze, Emmanuel Chukwudi, ed. *Race and the Enlightenment: A Reader*. Malden, MA: Blackwell Publishers, 1997.

Fee, Gordon D. "The Majority Text and the Original Text of the New Testament." In *Studies in the Theory and Method of New Testament Textual Criticism*, edited by Eldon Jay Epp and Gorden D. Fee, 107–18. Studies and Documents 45. Grand Rapids, MI: Eerdmans, 1993.

Felsenstein, Joseph. "Confidence Limits on Phylogenies: An Approach Using the Bootstrap." *Evolution* 39 (1985): 783–91.

Fisher, Ronald A. *The Genetical Theory of Natural Selection*. Oxford: Clarendon Press, 1930.

Foucault, Michel. *The Order of Things: An Archaeology of the Human Sciences*. New York: Vintage Books, 1973.

Fox, Adam. *John Mill and Richard Bentley: A Study of Textual Criticism of the New Testament 1675–1729*. Oxford: Blackwell, 1954.

Fuller, David Otis, ed. *True or False? The Westcott-Hort Textual Theory Examined*. Grand Rapids, MI: Grand Rapids International, 1973.

Galton, David. "Did Darwin Read Mendel?" *Quarterly Journal of Medicine* 102 (2009): 587–9.

Gamble, Harry Y. *Books and Readers in the Early Church: A History of Early Christian Texts*. New Haven, CT: Yale University Press, 1995.

Gee, Henry. *In Search of Deep Time: Beyond the Fossil Record to a New History of Life.* New York: Free Press, 1999.

Glansdorff, Nicolas, Ying Xu, and Bernard Labedan, "The Last Universal Common Ancestor: Emergence, Constitution and Genetic Legacy of an Elusive Forerunner." *Biology Direct* 3 (2008). Accessed April 24, 2013. http://www.biology-direct.com/content/3/1/29.

Gobineau, Joseph Arthur Comte de. *Essai sur l'inégalité des races humaines.* Paris: Didot, 1853.

———. *The Inequality of the Human Races.* Translated by Adrian Collins. New York: Putnam, 1915.

Grafton, Anthony. *Joseph Scaliger: A Study in the History of Classical Scholarship.* Vol. 1, *Textual Criticism and Exegesis.* Oxford: Oxford University Press, 1983.

Green, Zeth, and David C. Parker. "The Digital Nestle-Aland." Presentation given in the session "Critical Editions of the German Bible Society: Nestle-Aland 28th Edition," Society of Biblical Literature Annual Meeting, Chicago, November 19, 2012.

Haines-Eitzen, Kim. *Guardians of Letters: Literacy, Power, and the Transmitters of Early Christian Literature.* Oxford: Oxford University Press, 2000.

Haldane, John B. S. *A Mathematical Theory of Natural and Artificial Selection.* Cambridge: Cambridge University Press, 1924.

Hanna, Ralph. "The Application of Thought to Textual Criticism in All Modes—with Apologies to A. E. Housman." *Studies in Bibliography* 53 (2000): 163–72.

Haraway, Donna. "A Cyborg Manifesto: Science, Technology, and Socialist-Feminism in the Late Twentieth Century." In *Simians, Cyborgs, and Women: The Reinvention of Nature,* 149–81. New York: Routledge, 1991.

Hardy, Godfrey H. "Mendelian Proportions in a Mixed Population." *Science* 28 (1908): 49–50.

Harris, William H. *Ancient Literacy.* Cambridge, MA: Harvard University Press, 1989.

Head, Peter M., and Tommy Wasserman, eds. *Evangelical Text Criticism.* Accessed February 26, 2014. http://evangelicaltextualcriticism.blogspot.com/.

Henig, Robin Marantz. *A Monk and Two Peas: The Story of Gregor Mendel and the Discovery of Genetics.* London: Phoenix, 2001.

Hennig, Willi. *Grundzüge einer Theorie der phylogenetischen Systematik.* Berlin: Deutscher Zentralverlag, 1950.

———. *Phylogenetic Systematics.* Translated by D. Dwight Davis and Rainer Zangerl. Urbana, IL: University of Illinois Press, 1966.

Hills, Edward F. *The King James Version Defended! A Christian View of the New Testament Manuscripts.* Des Moines, IA: Christian Research, 1956.

Hodges, Zane C. "The Greek Text of the King James Version." In *Which Bible?* ed. David Otis Fuller, 25–38. Grand Rapids, MI: Grand Rapids International Publications, 1970.

Hoenigswald, Henry M., and Linda F. Wiener, eds. *Biological Metaphor and Cladistic Classification: An Interdisciplinary Perspective.* Philadelphia: University of Pennsylvania Press, 1987.

Holmes, Michael W. "What Text Is Being Edited? The Editing of the New Testament." In *Editing the Bible: Assessing the Task Past and Present*, ed. John S. Kloppenborg and Judith H. Newman, 91–122. Resources for Biblical Literature 69. Atlanta: Society of Biblical Literature, 2012.

———. "From 'Original Text' to 'Initial Text': The Traditional Goal of New Testament Textual Criticism in Contemporary Discussion." In *The Text of the New Testament in Contemporary Research: Essays on the Status Quaestionis*, edited by Bart D. Ehrman and Michael W. Holmes, 637–88. 2nd ed. New Testament Tools, Studies and Documents 42. Leiden: Brill, 2013.

Hort, Fenton John Anthony. *Life and Letters of Fenton John Anthony Hort*. Edited by Arthur Fenton Hort. London: Macmillan, 1896.

Housman, A. E., ed. *M. Manilii Astronomicon liber primus*. Cambridge: Cambridge University Press, 1903.

Howe, Christopher J., Adrian C. Barbrook, Linne Mooney, and Peter Robinson, "Parallels between Stemmatology and Phylogenetics." In *Studies in Stemmatology II*, edited by Pieter van Reenen, August den Hollander, and Margot van Mulken, 3–11. Amsterdam: John Benjamins, 2004.

Howe, Christopher J., Adrian C. Barbrook, Matthew Spencer, Peter Robinson, Barbara Bordalejo, and Linne R. Mooney. "Manuscript Evolution." *Trends in Genetics* 17 (2001): 147–52.

Hulley, Karl Kelchner. "Principles of Textual Criticism Known to St. Jerome." *Harvard Studies in Classical Philology* 55 (1944): 87–109.

Hume, David. "Of National Characters." In *Essays Moral and Political*, part I, sect. 21. Rev. ed. London: Millar, 1748.

Hunter, John. "Of Skinning." In *A Treatise on the Blood, Inflammation, and Gunshot Wounds*, 470–4. Philadelphia: Haswell, Barrington, and Haswell, (1794) 1840.

Hurtado, Larry W. "Beyond the Interlude? Developments and Directions in New Testament Textual Criticism." In *Studies in the Early Text of the Gospels and Acts: The Papers of the First Birmingham Colloquium on the Textual Criticism of the New Testament*, edited by David G. K. Taylor, 26–48. Texts and Studies 1. Atlanta: Society of Biblical Literature, 1999.

———. *The Earliest Christian Artifacts: Manuscripts and Christian Origins*. Grand Rapids, MI: Eerdmans, 2006.

Huson, Daniel H. "SplitsTree: Analyzing and Visualizing Evolutionary Data." *Bioinformatics* 14 (1998): 68–73.

Hyde, Lewis. *The Gift: Imagination and the Erotic Life of Property*. New York: Random House, 1979.

Isaac, Benjamin. *The Invention of Racism in Classical Antiquity*. Princeton: Princeton University Press, 2004.

James, Frank A. J. L. "An 'Open Clash between Science and the Church'? Wilberforce, Huxley, and Hooker on Darwin at the British Association, Oxford 1860." In *Science and Beliefs: From Natural Philosophy to Natural Science 1700–1900*, edited by David M. Knight and Matthew D. Eddy, 171–94. Aldershot, UK: Ashgate, 2005.

Jones, Alex. "Properties of a Stemma: Relating the Manuscripts in Two Texts from *The Canterbury Tales*." *Parergon* 18 (2001): 35–53.

Kannaday, Wayne C. *Apologetic Discourse and the Scribal Tradition: Evidence of the Influence of Apologetic Interests on the Text of the Canonical Gospels*. Leiden: Brill, 2004.

Kant, Immanuel. *Observations on the Feeling of the Beautiful and Sublime*. Translated by John T. Goldthwait. Berkeley: University of California Press, 1960.

———. "Von den verschiedenen Racen der Menschen." In *Immanuel Kant: Hauptwerke*, vol. 2, *Vorkritische Schriften 1757–1777*, edited by the Königlich-Preussische Akademie der Wissenschaften zu Berlin, 427–43. Charlottesville, VA: InteLex, (1775) 1999.

Kaplan, Gregory B. "The Inception of *Limpieza de Sangre* (Purity of Blood) and Its Impact in Medieval and Golden Age Spain." In *Marginal Voices: Studies in* Converso *Literature of Medieval and Golden Age Spain*, edited by Amy Aronson-Friedman and Gregory B. Kaplan, 19–41. Leiden: Brill, 2012.

Kelley, Shawn. *Racializing Jesus: Race, Ideology and the Formation of Modern Biblical Scholarship* (Biblical Limits). New York: Routledge, 2002.

———. "Race, Aesthetics, and Gospel Scholarship: Embracing and Subverting the Aesthetic Ideology." In *Prejudice and Christian Beginnings: Investigating Race, Gender, and Ethnicity in Early Christian Studies*, edited by Laura Nasrallah and Elisabeth Schüssler Fiorenza, 191–209. Minneapolis, MN: Fortress Press, 2009.

Kenyon, Frederic G. *Books and Readers in Ancient Greece and Rome*. Oxford: Clarendon Press, 1932.

———. *Recent Developments in the Textual Criticism of the Greek Bible*. Schweich Lectures of the British Academy, 1932. London: Oxford University Press, 1933.

Kloppenborg, John S., and Judith H. Newman, eds. *Editing the Bible*. Resources for Biblical Literature 69. Atlanta: Society of Biblical Literature, 2012.

Koshy, Susan. *Sexual Naturalization: Asian Americans and Miscegenation*. Stanford: Stanford University Press, 2004.

Lachmann, Karl, ed. *Sextus Aurelius Propertius: Carmina*. Leipzig: Fleischer, 1816.

———. "Rechenschaft über seine Ausgabe des Neuen Testaments." *Theologische Studien und Kritiken* 3 (1830): 817–45.

———. *Novum Testamentum Graece*. Vol. 1. Berlin: Reimer, 1831.

———. *Novum Testamentum Graece et Latine*. Vol. 2. Berlin: Reimer, 1850.

Lake, Kirsopp, Robert Blake, and Silva New. "The Caesarean Text of the Gospel of Mark." *Harvard Theological Review* 21 (October 1928): 206–404.

Lander, Eric S., Lauren M. Linton, Bruce Birren, et al. "Initial Sequencing and Analysis of the Human Genome." *Nature* 409 (2001): 860–921.

Lecointre, Guillaume, and Hervé Le Guyader. *The Tree of Life: A Phylogenetic Classification*. Translated by Karen McCoy. Cambridge: Harvard University Press, 2006.

Lee, Arthur R. "Numerical Taxonomy Revisited: John Griffith, Cladistic Analysis and St. Augustine's *Quaestiones in Heptateuchum*." In *Studia Patristica Vol. XX: Papers Presented at the Tenth International Conference on Patristic Studies Held in Oxford 1987*, ed. Elizabeth A. Livingstone, 24–32. Leuven: Peeters, 1989.

Lewis, Martin W., and Kären E. Wigen. *The Myth of Continents: A Critique of Metageography*. Berkeley: University of California Press, 1997.

Lindley, John. *Ladies' Botany or A Familiar Introduction to the Study of the Natural System of Botany*. London: Ridgeway, 1834.

Linnaeus, Carl (Carl von Linné). S.v. "Quadrupedia, Anthropomorpha, Homo." *Systema Naturae*. 1st ed. Leiden: Theodore Haak, 1735.

———. *Systema Naturae*. 10th ed. Halle: Kurt, 1760.

Locke, John. *An Essay Concerning Human Understanding and Treatise on the Conduct of the Understanding*. Philadelphia: Kay, 1844.

Lovejoy, Arthur O. *The Great Chain of Being: A Study of the History of an Idea*. Cambridge, MA: Harvard University Press, 1933.

Lowe, Lisa. *Immigrant Acts: On Asian American Cultural Politics*. Durham, NC: Duke University Press, 1996.

Maastricht, Gerhard von, ed. *'H ΔΙΑΘΉΚΗ Novum Testamentum*. Amsterdam: Wetsten, 1711.

Macgillivray, William. *A Manual of Botany: Comprising Vegetable Anatomy and Physiology or The Structure and Function of Plants*. London: Scott, Webster, and Geary, 1840.

Maddison, David R., and Katja-S. Schulz, eds. *The Tree of Life Web Project*. Accessed April 23, 2013. http://tolweb.org.

Mayr, Ernst. *Systematics and the Origin of Species*. New York: Columbia University Press, 1942.

———. "Introduction." In *On the Origin of Species; A Facsimile of the First Edition*, by Charles Darwin, Cambridge, MA: Harvard University Press, 1964.

McGann, Jerome. *A Critique of Modern Textual Criticism*. Charlottesville, VA: University Press of Virginia, 1983, paperback ed., 1992.

———. *The Textual Condition*. Princeton: Princeton University Press, 1991.

Metzger, Bruce M. "The Caesarean Text of the Gospels." *Journal of Biblical Literature* 64 (December 1945): 457–89.

———. *Chapters in the History of New Testament Textual Criticism*. Grand Rapids, MI: Eerdmans, 1963.

———. "Explicit References in the Works of Origen to Variant Readings in New Testament Manuscripts." In *Biblical and Patristic Studies: In Memory of Robert Pierce Casey*, edited by J. Neville Birdsall and Robert W. Thomson, 78–95. Freiburg: Herder, 1963.

Metzger, Bruce M., and Bart D. Ehrman. *The Text of the New Testament: Its Transmission, Corruption, and Restoration*. 4th ed. New York: Oxford University Press, 2005.

Michaelis, Johann David. *Einleitung in die göttlichen Schriften des Neuen Bundes*. Vol. 1. 2nd ed. Göttingen: Vandenhoeck, 1765.

Michigan News. "U-M Biologist Plays Key Role in Effort to Create First Comprehensive Tree of Life." *University of Michigan News Service*. May 21, 2012. Accessed April 23, 2013. http://ns.umich.edu/new/books-list/20400-u-m-biologist-plays-key-role-in-effort-to-create-first-comprehensive-tree-of-life.

Mill, John. *'Η ΚΑΙΝΗ ΔΙΑΘΗΚΗ Novum Testamentum*. Oxford: Sheldonian, 1707.

Mink, Gerd. "Editing and Genealogical Studies: The New Testament." *Literary and Linguistic Computing* 15 (2000): 51–6.

———. "Problems of a Highly Contaminated Tradition: The New Testament; Stemmata of Variants as a Source of Genealogy for Witnesses." In *Studies in Stemmatology II*, edited by Pieter van Reenan, August den Hollander, and Margot van Mulken, 13–86. Amsterdam: John Benjamins, 2004.

———. "The Coherence-Based Genealogical Method (CBGM)." Downloadable presentation available at the website of the University of Münster Institute for New Testament Münster: Textual Research. University of Münster Institute for New Testament Textual Research, 2009. Accessed May 14, 2013. http://egora.uni-muenster.de/intf/service/downloads_en.shtml.

———. "Contamination, Coherence, and Coincidence in Textual Transmission." In *The Textual History of the Greek New Testament: Changing Views in Contemporary Research*, edited by Klaus Wachtel and Michael W. Holmes, 141–216. Atlanta: Society of Biblical Literature, 2011.

———. "Guide to Genealogical Queries." Available at the website of the University of Münster Institute for New Testament Textual Research. Accessed May 15, 2013. http://intf.uni-muenster.de/cbgm/guide_en.html.

Miramon, Charles de. "Noble Dogs, Noble Blood: The Invention of the Concept of Race in the Late Middle Ages." In *The Origins of Racism in the West*, edited by Miriam Eliav-Feldon, Benjamin Isaac, and Joseph Ziegler, 200–16. New York: Cambridge University Press, 2009.

Moore, Stephen D., and Yvonne Sherwood. "Biblical Studies 'after' Theory: Onwards Towards the Past, Part One: After 'after Theory,' and Other Apocalyptic Conceits." *Biblical Interpretation* 18 (2001): 1–27.

Morgan, Thomas Hunt, Hermann Joseph Muller, and Alfred Henry Sturtevant. *The Mechanism of Mendelian Heredity*. New York: Holt, 1926.

Nordenskiöld, Erik. *The History of Biology: A Survey*. Translated by L. Eyre. London: Kegan Paul, Trench, Trubner, 1929.

Northrop, Filmer Stuart Cuckow. *The Meeting of East and West*. New York: Macmillan, 1960.

Ochman, Howard, Jeffrey G. Lawrence, and Eduardo A. Groisman, "Lateral Gene Transfer and the Nature of Bacterial Innovation," *Nature* 405 (2000): 299–304

O'Hara, Robert, and Peter Robinson. "Computer-Assisted Methods of Stemmatic Analysis." In *The Canterbury Tales Project Occasional Papers*, vol. 1, edited by Norman Blake and Peter Robinson, 53–74. Oxford: Office for Humanities Communication, 1993.

Olby, Robert, and Peter Gautrey. "Eleven References to Mendel before 1900." *Annals of Science* 24 (1968): 7–20

Pallen, Mark. *The Rough Guide to Evolution*. New York: Rough Guides, 2009.

Parker, David C. *Codex Bezae: An Early Christian Manuscript and Its Text*. Cambridge: Cambridge University Press, 1992.

———. Review of *The Orthodox Corruption of Scripture: The Effect of Early Christological Controversies on the Text of the New Testament* by Bart D. Ehrman. *Journal of Theological Studies* 45 (1994): 704–8.

———. *The Living Text of the Gospels*. Cambridge: Cambridge University Press, 1997.

———. "Manuscripts of the Gospels in the Electronic Age." *Restoration Quarterly* 42 (2000): 221–31.

———. "The Text of the New Testament and Computers: The International Greek New Testament Project." *Literary and Linguistics Computing* 15 (2000): 27–41.

———. "Through a Screen Darkly: Digital Texts and the New Testament." *Journal for the Study of the New Testament* 25 (2003): 395–411.

———. *An Introduction to the New Testament Manuscripts and Their Texts*. Cambridge: Cambridge University Press, 2008.

———. "*Et Incarnatus Est*." In *Manuscripts, Texts, Theology: Collected Papers 1977–2007*, 311–22. Berlin: de Gruyter, 2009.

———. *Textual Scholarship and the Making of the New Testament*. Oxford: Oxford University Press, 2012.

Pasquali, Giorgio. *Storia della tradizione e critica del testo*. 2nd ed. Florence: Le Monnier, 1962.

Pfeiffer, Rudolf. *History of Classical Scholarship: From the Beginnings to the End of the Hellenistic Age*. Oxford: Clarendon Press, 1968.

Philippe, Hervé, Henner Brinkmann, Dennis V. Lavrov, D. Timothy J. Littlewood, Michael Manuel, Gert Wörheide, and Denis Baurain. "Resolving Difficult Phylogenetic Questions: Why More Sequences Are Not Enough." *Public Library of Science (PLOS): Biology* 9 (2011). Accessed May 21, 2013. http://www.ncbi.nlm.nih.gov/pmc/articles/PMC3057953/.

Pichler, Alois. "Transcriptions, Texts and Interpretations." In *Culture and Value: Philosophy and the Cultural Sciences*, edited by Kjell Johannessen and Tore Nordenstam, 690–5. Vienna: Austrian Ludwig Wittgenstein Society, 1995.

Pick, Daniel. *Faces of Degeneration: A European Disorder c. 1848-c. 1924*. Ideas in Context 15. Cambridge: Cambridge University Press, 1989.

Powell, Baden. "On the Study of the Evidences of Christianity." In *Essays and Reviews: The 1860 Text and Its Reading*, edited by Victoria Shea and William Whitla, 233–73. Charlottesville, VA: University Press of Virginia, 2000.

Pray, Leslie A. "DNA Replication and Causes of Mutation." *Scitable* (2008). Accessed May 30, 2013. http://www.nature.com/scitable/topicpage/dna-replication-and-causes-of-mutation-409.

Quinet, Edgar. *Du Génie des Religions*. Paris: Charpentier, 1842.

Qureshi, Sadiah. "Displaying Sara Baartman, The 'Hottentot Venus.'" *History of Science* 42 (2004): 233–57.

Raven, Charles E. *John Ray Naturalist: His Life and Works*. Cambridge: Cambridge University Press, 1942.

Reeve, Michael. "Lucretius in the Middle Ages and Early Renaissance: Transmission and Scholarship." In *The Cambridge Companion to Lucretius*, edited by Stuart Gillespie and Philip Hardie, 205–13. Cambridge: Cambridge University Press, 2007.

Reynolds, Leighton D., and Nigel G. Wilson. *Scribes and Scholars: A Guide to the Transmission of Greek and Latin Literature*. 3rd ed. Oxford: Clarendon Press, 1991.

Roberts, Colin H. "Books in the Graeco-Roman World and in the New Testament." In *The Cambridge History of the Bible*, vol. 1, *From the Beginnings to Jerome*, edited by Peter R. Ackroyd and Christopher F. Evans, 48–66. Cambridge: Cambridge University Press, 1970.

Robinson, Maurice A., and William G. Pierpont. *The New Testament in the Original Greek According to the Byzantine/Majority Textform*. Atlanta: Original Word, 1991.

Robinson, Peter. "A Stemmatic Analysis of the Fifteenth-Century Witnesses to The Wife of Bath's Prologue." In *The Canterbury Tales Project Occasional Papers*, vol. 2, ed. Norman Blake and Peter Robinson, 69–132. Oxford: Office for Humanities Communication, 1997.

——— "New Methods of Editing, Exploring, and Reading *The Canterbury Tales*." Online article based on a talk at the conference "I nuovi orizzonti della filologia," Accademia Nazionale dei Lincei, Rome, May 28, 1998. Accessed July 22, 2015. http://www.petermwrobinson.me.uk/canterburytalesproject.com/pubs/desc2.html.

——— "What Text Really Is Not, and Why Editors Have to Learn to Swim." *Literary and Linguistic Computing* 24 (2009): 41–52.

Robinson, Peter, and Robert O'Hara. "Cladistic Analysis of a Norse Manuscript Tradition." In *Research in Humanities Computing*, vol. 4, ed. Susan Hockey and Nancy Ide, 115–37. Oxford: Oxford University Press, 1996.

Royse, James R. *Scribal Habits in Early Greek Papyri*. New Testament Tools, Studies, and Documents 36. Leiden: Brill, 2008.

Said, Edward. *Orientalism*. New York: Vintage Books, 1978.

Scaliger, Joseph. *Epistolae*. Leiden: Bonaventura and Abraham Elzevir, 1627.

Scholz, Johann Martin Augustin. *Novum Testamentum Graece*. 2 vols. Leipzig: Fleischer, 1830–6.

Segovia, Fernando. *Decolonizing Biblical Scholarship: A View from the Margins*. London: Orbis Books, 2000.

Semler, Johannes Salomo. "Vorrede." In *Hermeneutisches Vorbereitung*, vol. 3, part 1. Halle: Hemmerde, 1765.

——— *Apparatus ad liberalem Novi Testamenti*. Halle: Tramp, 1767.

Shea, Victoria, and William Whitla, eds. *Essays and Reviews: The 1860 Text and Its Reading*. Charlottesville, VA: University Press of Virginia, 2000.

Sicroff, Albert A. *Les controverses des status de "pureté de sang" en Espagne du XVe au XVIIe siècle*. Paris: Didier, 1960.

Slaughter, Mary M. *Universal Languages and Scientific Taxonomy in the Seventeenth Century*. Cambridge: Cambridge University Press, 1982.

Smith, Anthony. *The Ethnic Origin of Nations*. Oxford: Blackwell, 1986.

Smith, John Maynard. "Analyzing the Mosaic Structure of Genes." *Journal of Molecular Evolution* 34 (1992): 126–9.

Smith, Jonathan Z. *Drudgery Divine: On the Comparison of Early Christianities and the Religions of Late Antiquity*. Chicago: University of Chicago Press, 1990.

Spencer, Matthew, Elizabeth A. Davidson, Adrian C. Barbrook, and Christopher J. Howe. "Phylogenetics of Artificial Manuscripts." *Journal of Theoretical Biology* 227 (2004): 503–11.

Spencer, Matthew, Klaus Wachtel, and Christopher J. Howe. "The Greek Vorlage of the Syra Harclensis: A Comparative Study on Method in Exploring Textual Genealogy." *TC: A Journal of Biblical Textual Criticism* 7 (2002). Accessed May 16, 2013. http://rosetta.reltech.org/TC/v07/SWH2002/.

Spivak, Gayatri Chakravorty. *The Spivak Reader: Selected Works of Gayatri Chakravorty Spivak*. Edited by Donna Landry and Gerald MacLean. New York: Routledge, 1996.

Stolz, Michael. "New Philology and New Phylogeny: Aspects of a Critical Electronic Edition of Wolfram's *Parzival*." *Literary and Linguistic Computing* 18 (2003): 139–50.

Streeter, Burnett Hillman. *The Four Gospels: A Study of Origins*. London: Macmillan, 1924.

Strutwolf, Holger. "Scribal Practices and the Transmission of Biblical Texts." In *Editing the Bible: Assessing the Task Past and Present*, edited by John S. Kloppenborg and Judith H. Newman, 139–60. Resources for Biblical Literature 69. Atlanta: Society of Biblical Literature, 2012.

Sturtevant, Alfred H. *A History of Genetics*. New York: Harper & Row, 1965.

Sturtz, Harry A. *The Byzantine Text-Type and New Testament Textual Criticism*. Nashville, TN: Nelson, 1984.

Stuurman, Siep. "Francois Bernier and the Invention of Racial Classification." *History Workshop Journal* 50 (2000): 1–21.

Sullivan, John Patrick. *Martial: The Unexpected Classic*. Cambridge: Cambridge University Press, 1991.

Talmon, Shemaryahu. "The Old Testament Text." In *The Cambridge History of the Bible*, vol. 1, *From the Beginnings to Jerome*, edited by Peter R. Ackroyd and Christopher F. Evans, 159–99. Cambridge: Cambridge University Press, 1970.

Tanselle, G. Thomas. "Textual Criticism at the Millennium." *Studies in Bibliography* 54 (2001): 1–80.

Thomson, Keith. "Huxley, Wilberforce, and the Oxford Museum." *American Scientist* 88 (2000): 210–3.

Timpanaro, Sebastiano. *The Genesis of Lachmann's Method*. Edited and translated by Glenn W. Most. Chicago: University of Chicago Press, 2005.

Tregelles, Samuel Prideaux. *An Account of the Printed Text of the Greek New Testament*. London: Bagster, 1854.

Tyndale House Publishers. *The Living Bible, Paraphrased.* Edited by Kenneth N. Taylor. Wheaton, IL: Tyndale House, 1971.

——— *The Catholic Living Bible, Paraphrased: A Thought-for-Thought Translation.* Edited by Kenneth N. Taylor. Wheaton, IL: Tyndale House, 1976.

——— *New Living Translation.* Wheaton, IL: Tyndale House, 1996.

University of Münster Institute for New Testament Textual Research. "The *Editio Critica Maior.*" Website of the University of Münster Institute for New Testament Textual Research. Accessed June 5, 2013. http://www.uni-muenster.de/INTF/ECM.html.

Vincent, Marvin R. *A History of the Textual Criticism of the New Testament.* London: Macmillan, 1899.

Wachtel, Klaus. "The Coherence-Based Genealogical Method: A New Way to Reconstruct the Text of the Greek New Testament." In *Editing the Bible: Assessing the Task Past and Present*, ed. John S. Kloppenborg and Judith H. Newman, 123–9. Resources for Biblical Literature 69. Atlanta: Society of Biblical Literature, 2012.

——— "The Revision of the Catholic Epistles according to the *Editio Critica Maior.*" Presentation given in the session "Critical Editions of the German Bible Society: Nestle-Aland 28th Edition," at the Society of Biblical Literature Annual Meeting in Chicago, Illinois, November 19, 2012.

Wald, Priscilla. "Terms of Assimilation: Legislating Subjectivity in the Emerging Nation." In *Cultures of United States Imperialism*, edited by Amy Kaplan and Donald E. Pease, 59–84. Durham, NC: Duke University Press, 1993.

Wasserman, Tommy. "Criteria for Evaluating Readings in New Testament Textual Criticism," in *The Text of the New Testament in Contemporary Research: Essays on the Status Quaestionis*, edited by Bart D. Ehrman and Michael W. Holmes, 579–612. 2nd ed. New Testament Tools, Studies and Documents 42. Leiden: Brill, 2013.

Watson, James D. *The Double Helix: A Personal Account of the Structure of DNA.* New York: New American Library, 1969.

Wenzel, Siegfried. "Reflections on (New) Philology." *Speculum* 65 (1990): 11–8.

Westcott, Brook Foss. *Life and Letters of Brooke Foss Westcott.* 2 vols. Edited by Arthur Westcott. London: Macmillan, 1903.

Westcott, Brooke Foss, and Fenton John Anthony Hort, eds. *The New Testament in the Original Greek.* Cambridge: Macmillan, 1882.

Wexler, Barbara. *Genetics and Genetic Engineering.* Information Plus Reference Series. Detroit: ThompsonGale, 2011.

Wilberforce, Samuel. "On the Origin of Species." *Quarterly Review* 108 (1860): 225–64.

Wilkins, John. *An Essay Towards a Real Character, and a Philosophical Language.* London: Gellibrand, 1668.

Wilkins, John S. *Species: A History of the Idea.* Berkeley: University of California Press, 2009.

Williams, Charles S. C. *Alterations to the Synoptic Gospels and Acts.* Oxford: Blackwell, 1951.

Windram, Heather F., Christopher J. Howe, and Matthew Spencer. "The Identification in Exemplar Change in the *Wife of Bath's Prologue* Using the Maximum Chi-Squared Method." *Literary and Linguistic Computing* 20 (2005): 189–204.

Windram, Heather F., Prue Shaw, Peter Robinson, and Christopher J. Howe. "Dante's *Monarchia* as a Test Case for the Use of Phylogenetic Methods in Stemmatic Analysis." *Literary and Linguistic Computing* 23 (2008): 443–63.

Woese, Carl. "The Universal Ancestor." *Proceedings of the National Academy of Sciences* 95 (1998): 6854–9

Wright, Sewall. *Evolution and the Genetics of Populations*. 4 vols. Chicago: University of Chicago Press, 1968–78.

Index